Smart Ship Technology 2018

London, United Kingdom
23 - 24 January 2018

ISBN: 978-1-5108-8321-5

Printed from e-media with permission by:

Curran Associates, Inc.
57 Morehouse Lane
Red Hook, NY 12571

Some format issues inherent in the e-media version may also appear in this print version.

Printed by Curran Associates, Inc. (2019)

For permission requests, please contact The Royal Institution of Naval Architects
at the address below.

The Royal Institution of Naval Architects
8-9 Northumberland Street
London, WC2N 5DA
United Kingdom

Phone: 44 207 235 4622
Fax: 44 207 259 5912

publications@rina.org.uk

Additional copies of this publication are available from:

Curran Associates, Inc.
57 Morehouse Lane
Red Hook, NY 12571 USA
Phone: 845-758-0400
Fax: 845-758-2633
Email: curran@proceedings.com
Web: www.proceedings.com

CONTENTS

Smart Ship Technology

23 - 24 January 2018, London, UK

PAPERS

THE ROYAL INSTITUTION OF NAVAL ARCHITECTS
8-9 Northumberland Street, London, WC2N 5DA, Telephone: +44 (0)20 7235 4622

Exploiting machine learning for ship systems anomaly detection and healthiness forecasting

C. Gkerekos and **I. Lazakis** and **G. Theotokatos**, University of Strathclyde, UK

SUMMARY

This paper describes a novel methodology concerning the application of machine learning for the intelligent monitoring of ship systems. Monitoring of machinery condition is a crucial aspect of maintenance optimisation that is required for the vessel operation to remain sustainable and profitable. Operator-approved performance data are used to train a data-driven Expected Behaviour Model (EBM). Once trained, this model accepts newly-acquired data points as input and returns the probability of them belonging to the same performance profile with training data. Through this, emerging anomalies can be detected. This tool is coupled with a short-term healthiness forecasting tool, able to estimate future healthiness index. This combination allows the derivation of a healthiness score of both current and future condition of ship machinery. Additionally, in cases of performance degradation, the Remaining Useful Life (RUL) of given system can be projected. This provides a robust framework for the early detection of incipient machinery faults.

1. INTRODUCTION

Ships are a crucial asset of the global goods transportation system, carrying over four-fifths of merchandise. Combining current financial situation of the shipping industry with an average global merchant fleet vessel age of almost twenty years [16] successfully conveys the need of a high level of operation optimisation in order for the vessel to remain sustainable and profitable.

Maintenance of ship's machinery components can substantially affect its sustainability and profitability. Meanwhile, the gap between current maintenance state-of-art and ship maintenance state-of-practice leaves much to be desired. For this reason, the introduction of novel approaches of condition-monitoring, tailored to shipping industry's requirements and peculiarities is crucial. These approaches ought to provide a streamlined path from data acquisition to maintenance decision making in order to introduce a low threshold of adoption without majorly altering the established modus operandi of shipping companies. At the same time, these solutions should be easily modifiable in order to suit specific requirements while retaining a degree of openness to the end-user, so that results and methodology can be compared to common judgment and evaluated in order to eventually be trusted.

This research paper aims to introduce a novel methodology concerning the application of machine learning for the intelligent monitoring of ship systems. Operator-approved performance data are used to train a data-driven expected behaviour model based on Support Vector Machines (SVMs). Once trained, this model accepts newly-acquired data points as input and returns the probability of them belonging to the same performance profile with training data. Through this, emerging anomalies can be detected. This tool is coupled with a short-term healthiness forecasting tool, able to estimate future healthiness index. This combination allows the derivation of a healthiness score of both current and future condition of ship machinery. Additionally, in cases of performance degradation, the Remaining Useful Life (RUL) of given system can be projected. This will provide a robust framework for the early detection of incipient machinery faults.

Section 1 introduces the paper's scope and motivation of research. Section 2 reviews current research literature, identifying key points. Section 3 elaborates on the proposed methodology and Section 4 details the set-up of a case study used to validate proposed methodology. Section 5 presents and discusses the results obtained through the case study of Section 4, Finally, in Section 6 overall conclusions are provided along with future research steps.

2. RESEARCH BACKGROUND

This section provides an overview of literature pertinent to this paper. First, maintenance strategies are compared, followed by a review of how maintenance is regarded in the maritime sector. Besides, methodologies relevant to performance monitoring of machinery are included.

2.1 MAINTENANCE STRATEGIES

Four main maintenance types exist: reactive, preventive, predictive and proactive. Reactive maintenance concerns maintenance that is only performed once a component fails completely. Preventive maintenance refers to maintenance that happens at a fixed frequency (also known as PMS - Planned Maintenance System). Predictive maintenance utilises an estimation of the current and future condition of a component in order to obtain to provide an optimised maintenance scheduling that prevents failures without resorting to over-maintenance. This optimised maintenance scheduling offers extended machine lifespan, coupled with reduced maintenance costs and downtime. Proactive maintenance can be considered an extension of predictive maintenance, identifying the underlying causes of a fault following its detection.

2.1 MAINTENANCE IN THE MARITIME SECTOR

In sectors such as defence, aviation, manufacturing, automobile, and nuclear power production, maintenance focus has shifted from reactive to preventive/ predictive. Ship maintenance amounts to 10-15% of the shipping company direct operating costs [15].

However, in the maritime sector, ship maintenance has been considered an area of needless expenditure and advanced monitoring methods have not yet been widely applied [11]. Nevertheless, some attempts towards predictive maintenance in shipping have been made in the past few years. For example, a methodology where vibration data are combined with performance data (cylinder pressures) for the condition monitoring of a main engine has been suggested [1]. Accordingly, a thermodynamic model of a main engine has been developed to perform condition monitoring using cylinder pressure traces [17]. Besides, a self-learning algorithm for fault diagnosis in the combustion system of a marine diesel engine has been developed [12].

Lazakis et al. [10] present the framework of INCASS project, an EU FP7 project that developed tools to enhance machinery monitoring by combining real time information with machinery risk analysis tools. In this scope, a machinery risk analysis tool that performs condition monitoring and maintenance decision support was suggested. Dikis et al. [3] further elaborate this tool. Furthermore, Gkerekos et al. [6] developed a self-learning model for the condition monitoring of ship machinery based on vibration measurements. Accordingly, Gkerekos et al. [7] developed a self-learning model for condition monitoring of machinery components using raw physical data collected through measuring campaigns on board vessels.

2.3 MACHINERY PERFORMANCE MONITORING

Performance monitoring of machinery is a problem that requires the development of a suitable model. This model can either use a first-principles analysis (i.e. white-box model), or use a more "brute-force" approach by developing a model using self-learning algorithms coupled with an acquired dataset (i.e. black-box model). Often, a combination of both techniques is applied, leading to grey-box models. Yang et al. [19] developed a framework for the analysis of data acquired through wind turbine SCADA (Supervisory Control and Data Acquisition) bus measurements to perform condition monitoring based on correlations between measurements. Widodo & Yang [18] presented an overview of Support Vector Machine (SVM) techniques for fault diagnosis and monitoring in engineering applications. Yin & Hou [20] built on that review, providing an updated overview of SVM applications for fault diagnosis and process monitoring.

3. SUGGESTED METHODOLOGY

In this section, the suggested methodology for this paper is elaborated, in a module-by-module fashion. The modules included contain the necessary functionality to perform pre-processing, anomaly detection, healthiness estimation, and healthiness forecasting. Each will be further elaborated in the following subsections.

A visual representation of the suggested methodology is presented in Figure 1, presenting all suggested modules and their relevant interconnections.

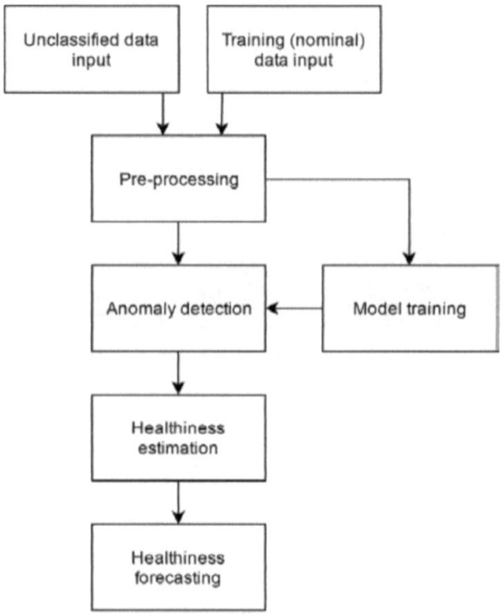

Figure 1: Visual representation of the suggested methodology.

3.1 PRE-PROCESSING

Elements of the dataset that are missing are replaced by "Not a Number" (NaN), a notation used to denote non-numerical values in cases where a numerical value is expected. The pre-processing methodology is elaborated below:

- Rows containing NaN values are dropped.
- Features are scanned for elements with values beyond $\mu \pm 3\sigma$ and corresponding rows are dropped. μ corresponds to the mean value of each column and σ to its standard deviation.
- New features are derived in order to better present actual condition. For example, in cases where e.g. temperature or pressure inlet and outlet values are available, their differential is calculated and added to the dataset, replacing original measurements.
- Remaining dataset are split into training (70%), validation (10%), and testing (20%) sub-datasets.
- Elements μ and σ of training sub-dataset's features x_T are extracted and used to produce normalised sub-dataset x'_T as shown in Equation 1.

- Validation and testing sub-datasets are also normalised using μ and σ values extracted from training sub-dataset.

$$x'_T = \frac{x_T - \mu}{\sigma} \qquad (1)$$

While training sub-dataset is used for model parameter optimisation, validation sub-dataset is used to select the optimal hyperparameter values. Finally, testing sub-dataset is a previously-unseen dataset that is used to evaluate the generalisation error of the final selected model [8]. Whereas model parameters are training parameters that are learnt during model training, model hyperparameters cannot be learnt during training but instead are set in advance. Hence, while model parameters vary for distinct training data sets of the same application, hyperparameters tend to be common for similar models.

3.2 ANOMALY DETECTION

A one-class Support Vector Machine (SVM) classifier [2] was trained. Compared to other machine learning tools, SVMs offer superior generalisation capabilities [18]. This is a semi-supervised algorithm that learns a decision function for outlier detection. In other words, it classifies new observations as similar or different to the dataset used for training.

A key selection point in a SVM implementation concerns the kernel function to be used. Kernels operate as a similarity function, offering a gauge of similarity between two inputs. While multiple kernel functions can be applied for model training, Radial Basis Function (RBF) is considered to work well in practice and is usually a reasonable first choice [9]. A significant advantage of RBF compared to other kernel functions is that it can be easily calibrated as it only depends on one parameter, $\gamma \in (0, \infty)$. Parameter γ controls the region that a single training sample can affect, with a small value of γ increasing the size of this region; and conversely. In other words, a large value of γ leads towards over-fitting in the Bias-Variance trade-off [5] whereas selecting a small value risks creating a too constrained model that cannot efficiently capture the complexity of the data. For two vectors u and v, RBF function $\kappa(u, v)$ is shown in Equation 2.

$$\kappa(u, v) = \exp(-\gamma \|u - v\|^2) \qquad (2)$$

Apart from the selection of a kernel function and the corresponding parameters, performance of one-class SVM classifiers can be calibrated through a single parameter $v \in (0, 1]$ [14]. Parameter v presents an upper bound on the fraction of margin errors and a lower bound of the fraction of support vectors [13]

From the above, it is clear that the performance of a one-class SVM classifier utilising RBF as kernel function and given a specific training dataset depends on the two aforementioned hyperparameters: γ and v. The optimal values of these hyperparameters are selected through grid search, i.e. an exhaustive search through a manually selected subset of the hyperparameter space.

Following the above, $(0, 1]$ is the range considered for hyperparameter v. Extreme values inside this range are preferably avoided as they usually tend to direct the model towards either a large number of miss-classifications, or a small number of support vectors leading to problematic model generalisation. In the case of hyperparameter γ a subset of reasonable values has to be selected as a parameter sweep in $(0, \infty)$ is not feasible. Hsu et al. [9] suggest the use of exponentially growing sequences of γ, e.g. the range of $\gamma = 2^{-15}, 2^{-13}, \dots, 2^3$ in order to identify the best pair of (v, γ). This is complemented by cross-validation in order to reduce the risk of over-fitting.

Models are trained for each point of the two dimensional grid using the training sub-dataset and then their performance using validation dataset is recorded. The hyperparameters of model that achieves the best results against the validation dataset are finally selected.

3.3 HEALTHINESS ESTIMATION

In the "anomaly detection" step of subsection 3.2, a hyperplane that encompasses original healthy points is derived in order to separate accordingly new points into healthy and unhealthy. At the same time, the distance between hyperplane and each new point can be used as a healthiness metric. The furthest a point is from the hyperplane towards the healthy side, its healthiness increases, and conversely. This is then mapped into a 0-100% semi-qualitative range in order to simplify evaluation and further processing. The points utilised to perform this mapping are case-specific and, accordingly, user-selectable.

Additionally, as every data point is evaluated independently, a moving average filter is implemented to smooth-out unnecessary kinks and make the evaluation of the overall degradation easier.

3.4 HEALTHINESS FORECASTING

Finally, in this step healthiness of past points are used as a basis of short-term forecasting of future cases. This is beneficial for the early identification of incipient faults. Additionally, if such a degradation is detected, the Remaining Useful Life (RUL) of the system or component can be estimated.

For this, a suitable threshold must be set by the user. Forecasting works by implementing least-squares in order to identify the most accurate straight-line that describes the model output. Input for this is the waited average line, as this by definition, is smoother and can be more accurately be described by a straight-line, i.e.

$$y = A\,x + B$$

4. METHODOLOGY APPLICATION

In this section, a case study utilising Diesel Generator (D/Gen) data is presented to demonstrate the methodology implementation. Six features were selected for model training: as shown in Table 1. 6 observations were available per day, for a total of two months. As the specific DGen was not always operational, 191 points were acquired. In order to evaluate the performance of the suggested methodology, simulated fault patterns were introduced in the dataset in order to evaluate relevant anomaly detection performance.

Table 1. Performance measurements utilised as input for model training.

Component	Measurement description
Misc.	Power output (kW)
L.O.	Temperature (^oC)
	Inlet pressure (kg/cm^2)
F.W.	Cool. inlet (^oC)
Exhaust gas	Min, max values (^oC)

5. APPLICATION RESULTS

This section presents and discusses the results obtained through the case study presented above, in Section 4. This section will follow the visual representation of Figure 1, and elaborate the results obtained along the way, utilising each suggested module.

5.1 PRE-PROCESSING

The dataset used for this case study were obtained from a manually-filled engine logbook. Thus, some errors could have snuck in due to human error. For this reason, a pre-processing module is introduced, as described in Section 3 above.

In Fig. 2, actual points obtained for LO inlet pressure are shown with two lines denoting $\mu \pm 3\sigma$ superimposed on the graph. There, two points (one around sample number 20 and one around 290) are shown to deviate from these bands. Due to the relative stability of neighbouring points, one can safely assume that these points are outliers either due to sensor malfunction or mistyping and discard them.

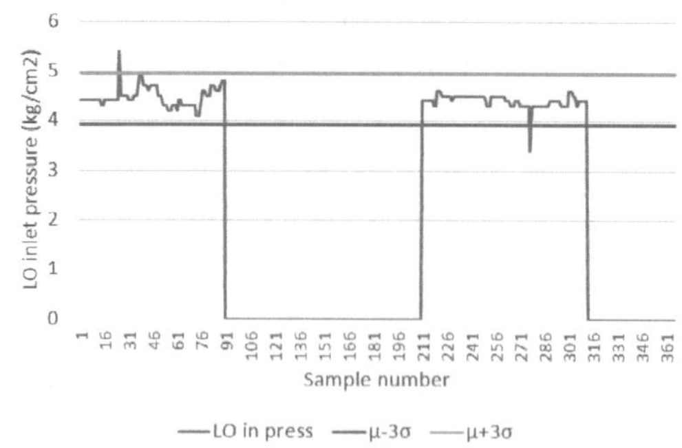

Figure 2: LO inlet pressure points with superimposed μ +/- 3σ.

5.2 ANOMALY DETECTION MODEL TRAINING

Grid search evaluating 1000 points was selected as this provides a good balance between grid resolution and computational cost. To reduce the risk of over-fitting the dataset, k-folding with k=10 was implemented in order to identify the most suitable combination of (ν, γ). The optimal values of these hyperparameters are presented in Table 2.

Table 2. Optimal hyperparameter values

hyperparameter	value
γ	2^(-5.11)
ν	0.029

5.3 HEALTHINESS ESTIMATION

In order to evaluate the healthiness estimation module, a simulated fault was introduced in the dataset. In this case, the dataset was replicated and the following changes were made:

- Power output was linearly reduced from 100% to 85%, (i.e., first point of the dataset was the only one left untouched and last point corresponded to 85% of the original, with all other points being manipulated accordingly).
- LO inlet pressure was linearly reduced from 100% to 80%.

Fig. 3 presents the results of the healthiness estimation. As every point is evaluated independently, there are some up and downs in the original (healthy) parts (e.g. around sample 75 and 190). For this reason, a smoothing filter (blue line) is introduced. Evaluating this line, a relatively flat line is obtained for points that are considered healthy, whereas a sharp drop is observed as the fault is introduced and then becomes full-scale.

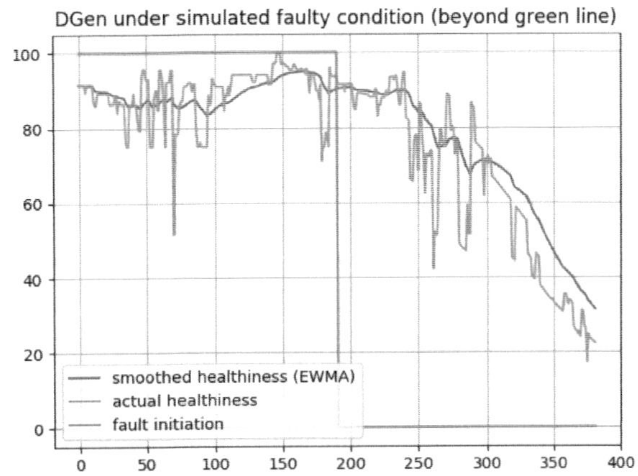

Figure 3: Healthiness estimation before (left of green line) and after (right of green line) introducing simulated fault.

5.4 HEALTHINESS FORECASTING

Here, the output of the moving average is used as input to a least-squares model that returns a straight-line that is the model forecast for future healthiness. This can be seen in Fig. 4 as the thick line at the end of the 'smoothed healthiness' line.

In order to obtain this, the last 70 points of the filtered healthiness estimation were used to train the least-squares model.

The optimal values that were identified are presented in Table 3 below.

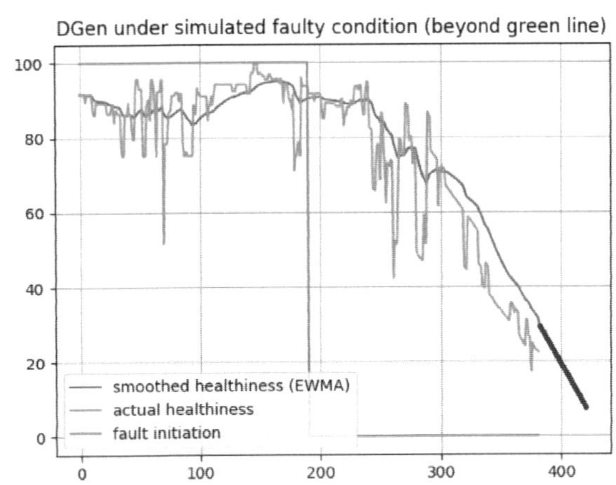

Figure 4: Healthiness estimation including straight-line forecasting (thick line).

Table 3. Optimal least-squares values for forecasting

hyperparameter	value
A	-0.58
B	29.60

6. CONCLUSIONS

This paper presents a robust but flexible framework for the pre-processing and analysis of performance data to monitor the condition of ship machinery through the extraction of a healthiness index. First, an overview of the current state of research in the field of maritime maintenance and condition monitoring is provided. Proposed methodology was elaborated and then showcased through a case study.

The case study included showed that a data-driven approach can accurately detect incipient faults in ship machinery using the vast amount of measurements that are collected through modern on-board data collection systems.

To conclude, future research steps include further model validation and integration with a Maintenance Decision Support System (M-DSS). Additionally, more advanced methods of forecasting will be implemented in order to obtain more accurate results.

7. ACKNOWLEDGEMENTS

The work presented in this paper is partially funded by Integrated Ship Energy & Maintenance Management System (ISEMMS) project. ISEMMS project has received research funding from Innovate UK under Project No. 102435. This publication reflects only the authors' views and Innovate UK is not liable for any use that may be made of the information contained within.

8. REFERENCES

1. Chandroth, G. (2004). Condition monitoring: the case for integrating data from independent sources. Journal of Marine Engineering & Technology 3(1), 9–16.
2. Cortes, C. & V. Vapnik (1995). Support-vector networks. Machine Learning 20(3), 273–297.
3. Dikis, K., I. Lazakis, A. Michala, Y. Raptodimos, & G. Theotokatos (2017). Dynamic risk and reliability assessment for ship machinery decision making. In Risk, Reliability and Safety: Innovating Theory and Practice - Proceedings of the 26th European Safety and Reliability Conference, ESREL 2016.
4. Dinwoodie, I. (2014). Modelling the Operation and Maintenance of Offshore Wind Farms. Ph. D. thesis, University of Strathclyde, Department of Electronic and Electrical Engineering.
5. Geman, S., E. Bienenstock, & R. Doursat (1992). Neural networks and the bias/variance dilemma. Neural computation 4(1), 1–58.
6. Gkerekos, C., I. Lazakis, & G. Theotokatos (2016, October). Ship machinery condition monitoring using vibration data through supervised learning. In I. Lazakis and G.

Theotokatos (Eds.), Proceedings of MSO2016, International Conference on Maritime Safety and Operations, pp. 103–110. University of Strathclyde Print Services.

7. Gkerekos, C., I. Lazakis, & G. Theotokatos (2017, January). Ship machinery condition monitoring using performance data through supervised learning. In Proceedings of Smart Ships Technology 2017 Conference, pp. 105–111. Royal Institution of Naval Architects (RINA).
8. Hastie, T., R. Tibshirani, & J. Friedman (2001). The Elements of Statistical Learning. Springer Series in Statistics. New York, NY, USA: Springer New York Inc.
9. Hsu, C.-W., C.-C. Chang, & C.-J. Lin (2010). A practical guide to support vector classification.
10. Lazakis, I., K. Dikis, A. L. Michala, & G. Theotokatos (2016). Advanced ship systems condition monitoring for enhanced inspection, maintenance and decision making in ship operations. Transportation Research Procedia 14, 1679 – 1688.
11. Lazakis, I. & A. Olcer (2015). Selection of the best maintenance approach in the maritime industry under fuzzy multiple attributive group decision-making environment. Proceedings of the Institution of Mechanical Engineers, Part M 230(2), 297–309.
12. Li, P., L. Liu, & H. Gong (2010). The research of the intelligent fault diagnosis optimized by aca for marine diesel engine. In Q. Luo (Ed.), Advancing Computing, Communication, Control and Management, pp. 174–181. Springer.
13. Pedregosa, F., G. Varoquaux, A. Gramfort, V. Michel, B. Thirion, O. Grisel, M. Blondel, P. Prettenhofer, R. Weiss, V. Dubourg, J. Vanderplas, A. Passos, D. Cournapeau, M. Brucher, M. Perrot, & E. Duchesnay (2011). Scikit-learn: Machine learning in Python. Journal of Machine Learning Research 12, 2825–2830.
14. Scholkopf, B., A. J. Smola, R. C. Williamson, & P. L. Bartlett (2000, May). New support vector algorithms. Neural Computation 12(5), 1207–1245.
15. Stopford, M. (2008). Maritime Economics (Third Edition ed.).Routledge.
16. UNCTAD (2015). Review of maritime transport. Technical report, United Nations Conference on Trade and Development,New York and Geneva.
17. Watzenig, D., M. S. Sommer, & G. Steiner (2009). Engine state monitoring and fault diagnosis of large marine diesel engines. Elektrotechnik und Informationstechnik 126(5), 173–179.
18. Widodo, A. & B.-S. Yang (2007). Support vector machine in machine condition monitoring and fault diagnosis. Mechanical Systems and Signal Processing 21(6), 2560 – 2574.
19. Yang, W., R. Court, & J. Jiang (2013). Wind turbine condition monitoring by the approach of SCADA data analysis. Renewable Energy 53, 365 – 376.
20. Yin, Z. & J. Hou (2016). Recent advances on SVM based fault diagnosis and process monitoring in complicated industrial processes. Neurocomputing 174, Part B, 643 – 650.

9. AUTHORS BIOGRAPHY

Christos Gkerekos is a PhD Researcher at the department of Naval Architecture, Ocean and Marine Engineering (NAOME) of University of Strathclyde. His research focuses on the use of self-learning algorithms for the condition monitoring of marine machinery.

Iraklis Lazakis is a Senior Lecturer at the department of NAOME of University of Strathclyde. His research interests include ship systems condition monitoring, systems maintenance and reliability, ship operations, risk analysis tools and methodologies, system criticality assessment, shipyard manufacturing and productivity, and asset management.

Gerasimos Theotokatos is a Lecturer at the department of NAOME of University of Strathclyde. His research interests include ship propulsion plant modelling, performance and emissions prediction of marine Diesel engines, the development and testing of propulsion plant system control schemes, the analysis of engine and its turbocharging system under normal and adverse operating conditions, and the experimental investigation of performance and emission parameters of internal combustion engines.

On-board Monitoring Data Analysis Based on Kernel Regression Model: Analysis of Shaft Power Component

M Minoura, Osaka University, Japan

SUMMARY

A statistical model for evaluating ship service performance with on-board monitoring data is proposed. A shaft power function, which is representative of service performance, is discussed. A statistical model is based on the kernel multivariate regression model. The dependent variables of the kernel regression model must be measureable; e.g. ship speed and motion. The simulated on-board monitoring data were provided to validate the regression model. When these data were transformed to be independent under the constraints of dynamic conditions, it became clear that the regression model is accurately identified and shaft power is rationally analysed. Dynamic conditions are derived from physical problems such as shaft power is zero when ship speed is zero. This regression model enables to decompose the shaft power to each cause.

1. INTRODUCTION

This paper addresses a method for estimating the shaft power function based on a statistical analysis of actual voyage data recorded on board automatically which is called on-board monitoring data.

It is useful from the viewpoint of energy saving operation of ships to know the components of the shaft power or the engine power. The resistance acting on the ship hull can generally be classified to some components like still water resistance, wind resistance, wave resistance. If we can clarify to what extent each component affects the engine power, it can contribute to improving the ship operation, and performance development targets can be clearly segmented. Therefore the author proposes the shaft power function for component analysis of shaft power.

A shaft power function is defined as a function with several variables expressed as ship responses. Selecting these variables, we can decompose monitored shaft power into components caused by still water resistance, wind resistance, wave resistance. A kernel multivariate regression model is applied as a shaft power function.

When on-board monitoring data are applied to a kernel multivariate regression model, they must be independent statistically. In addition, if constraint conditions based on a physical fact is imposed in a statistical model, model parameters can be identified accurately shaft power can be decomposed exactly. In this paper these matters are shown.

2. SHAFT POWER FUNCTION

2.1 SHORT-TERM SHIP RESPONSE

Sea-state such as waves, winds and ocean currents encountered by ships is stochastically steady during about 30 minutes. Statistics of waves, winds and ocean currents during this period are defined as short-term sea-state, and statistics such as ship motions, engine motions, and structural deformation are defined as short-term ship response. This statistic is expressed as the mean and variance of the instantaneous time series. The moving average and the moving variance are obtained as time series with different time scales. Further, the moving average and the moving variance of the statistic can be defined as new statistics. Sea-state and short-term ship response are classified as follows.

Sea-state:
Significant wave height, Mean wave period, Primary wave direction, Mean wind speed, Prevailing wind direction, etc.

Maneuvering system (mean):
Ship speed, Course, Drift angle, Heading, Draft, etc.

Engine system (mean):
Revolution speed, Torque, Thrust, Shaft power, Fuel consumption, etc.

Hull system (standard deviation):
Ship motion, Stress, Deformation, Wave load, etc.

From the viewpoint of the ship master, the maneuvering system is not the response amount, but it is considered as the response amount of the control system trying to satisfy the condition of the controlled object (course or engine revolution speed) with respect to the sea state. Such short-term ship response is to be analysed.

2.2 FUNCTIONAL DESCRIPTION OF SHAFT POWER

Let the shaft power be expressed as a function of short-term ship response. If the dominant variables are assumed to be the ship response such as ship speed, wind speed, ship motion, wave height, wave steepness, drift angle, draft, etc., the shaft power function is defined as

$$P_s = \hat{P}_s(V_{rcx}, V_{rax}, V_{ray}, H_{1/3}, G_{rx}, G_{ry}, X_{r4}, X_{r5}, \beta, D_m) + \varepsilon, \tag{1}$$

where,

P_s	Shaft power (mean)
$\hat{P}_s$	Statistical model of shaft power
ε	Residual
V_{rcx}	Long. relative speed for water (mean)
V_{rax}	Long. relative speed for air (mean)
V_{ray}	Trans. relative speed for air (mean)
$H_{1/3}$	Significant wave height
G_{rx}	Long. mean wave steepness
G_{ry}	Trans. mean wave steepness
X_{r4}	Roll rate (standard deviation)
X_{r5}	Pitch rate (standard deviation)
β	Drift angle (mean)
D_m	Draft (mean)

This function can decompose the shaft power: e.g. the shaft power for the draft D_m and the ship speed V_{rcx} in a calm sea condition without waves, winds and currents is

$$P_{s1} = \hat{P}_s(V_{rcx}, 0, \cdots, 0, D_m) + \varepsilon_1. \quad (2)$$

The shaft power in relative winds or waves is described as

$$P_{s2} = \hat{P}_s(V_{rcx}, V_{rax}, V_{ray}, 0, \cdots, 0, D_m) + \varepsilon_2, \quad (3)$$

$$P_{s3} = \hat{P}_s(V_{rcx}, 0, 0, H_{1/3}, G_{rx}, G_{ry}, X_{r4}, X_{r5}, 0, D_m) + \varepsilon_3. \quad (4)$$

If taking a subtraction of Eq. 2, we can estimate the added shaft power in relative winds or waves as

$$\Delta P_{s(wind)} = P_{s2} - P_{s1}, \quad (5)$$

$$\Delta P_{s(wave)} = P_{s3} - P_{s1}. \quad (6)$$

On the other hand, this estimation is also realised by reverse calculation excluding the target ship response. The shaft power excluding each target ship response is described as

$$P_{s4} = \hat{P}_s(V_{rcx}, 0, 0, H_{1/3}, G_{rx}, G_{ry}, X_{r4}, X_{r5}, \beta, D_m) + \varepsilon_3, \quad (7)$$

$$P_{s5} = \hat{P}_s(V_{rcx}, V_{rax}, V_{ray}, 0, \cdots, 0, \beta, D_m) + \varepsilon_3. \quad (8)$$

If taking a subtraction of Eq. 1, we can estimate the added shaft power in relative winds or waves as

$$\Delta P'_{s(wind)} = P_s - P_{s4}, \quad (9)$$

$$\Delta P'_{s(wave)} = P_s - P_{s5}. \quad (10)$$

If added shaft powers in disturbance are individually independent, the both calculation must be same.

$$\Delta P_{s(wind)} = \Delta P'_{s(wind)},$$

$$\Delta P_{s(wave)} = \Delta P'_{s(wave)}.$$

3. KERNEL REGRESSION MODEL

3.1 FORMULATION

A kernel multivariate regression model is applied as a statistical model for on-board monitoring data analysis. This model has a structure in which the variables of the linear regression model are replaced with kernel functions, and it is generalized as follows.

$$\hat{y}(\boldsymbol{x}) = \sum_{j=1}^{n} \alpha_j k(\boldsymbol{x}, \boldsymbol{x}^{(j)}), \quad (11)$$

where, variables and parameters as follows.

$\hat{y}$	Response variable (Dependent variable)
$\boldsymbol{x}$	Input vector variable (Independent variable) $\boldsymbol{x} = [x_1, x_2, \cdots, x_m]^T$
m	Dimension number of $\boldsymbol{x}$
$y^{(j)}$	Leaning data of response variable
$\boldsymbol{x}^{(j)}$	Leaning data of input variable $\boldsymbol{x}^{(j)} = \left[x_1^{(j)}, x_2^{(j)}, \cdots, x_m^{(j)}\right]^T$
j	Serial number of learning data $j = 1, 2, \cdots, n$
n	Length of learning data
α_j	Regression coefficient
$k(\boldsymbol{x}, \boldsymbol{x}^{(j)})$	Kernel function

Unlike the linear regression model, the number of terms is as same as that of learning data. The learning data are included as parameters in the kernel function. Therefore, the degree of freedom is extremely high as compared with the linear regression model. The Gaussian kernel and polynomial kernel are known well.

Gaussian kernel

$$k(\boldsymbol{x}, \boldsymbol{x}^{(j)}) = \exp\left(-\beta \|\boldsymbol{x} - \boldsymbol{x}^{(j)}\|^2\right) \text{ for } \beta > 0. \quad (12)$$

Polynomial kernel

$$k(\boldsymbol{x}, \boldsymbol{x}^{(j)}) = (\boldsymbol{x}^T \boldsymbol{x}^{(j)} - \mathrm{c})^p \text{ for } c \geq 0, p = 1, 2, \cdots. \quad (13)$$

The Gaussian kernel evaluates the closeness between learning data and variable as the term of $\|\boldsymbol{x} - \boldsymbol{x}^{(j)}\|$ shows. At the point that the variable agrees with the learning data, the kernel becomes the maximum 1. At the far field from the learning data, the kernel converges to zero.

The polynomial kernel diverges when the variable leaves the origin as the term of $\boldsymbol{x}^T \boldsymbol{x}^{(j)}$. This kernel is suitable for the matter that the response variable increases with respect to the input variables in the extrapolation area (outside the range of the learning data). It is naturally considered that the polynomial kernel is appropriate for the relationship of the shaft power to the ship speed. However, in principle it is impossible for the extrapolated region far apart. Hence, the kernel regression model must be applied in the interpolation region and a slightly further extrapolation region.

3.2 REGRESSION COEFFICIENT

A true response variable is described by the following equation using the kernel regression model.

$$y(\boldsymbol{x}) = \hat{y}(\boldsymbol{x}) + \varepsilon, \tag{14}$$

where, ε is residual. A square of residual is called the cost function defined as

$$R_0 = \sum_{j=1}^{n} \left\{ y^{(i)} - \sum_{j=1}^{n} \alpha_j k\left(\boldsymbol{x}^{(i)}, \boldsymbol{x}^{(j)}\right) \right\}^2, \tag{15}$$

By using the inner product of vectors, this expression can simplify as follows.

$$R_0(\boldsymbol{\alpha}) = (\boldsymbol{y}' - \boldsymbol{K\alpha})^T(\boldsymbol{y}' - \boldsymbol{K\alpha}), \tag{16}$$

where, each vector or matrix is defined as

$$\boldsymbol{K} = \begin{bmatrix} k\left(\boldsymbol{x}^{(1)}, \boldsymbol{x}^{(1)}\right) k\left(\boldsymbol{x}^{(1)}, \boldsymbol{x}^{(2)}\right) \dots k\left(\boldsymbol{x}^{(1)}, \boldsymbol{x}^{(n)}\right) \\ k\left(\boldsymbol{x}^{(2)}, \boldsymbol{x}^{(1)}\right) k\left(\boldsymbol{x}^{(2)}, \boldsymbol{x}^{(2)}\right) \cdots k\left(\boldsymbol{x}^{(2)}, \boldsymbol{x}^{(n)}\right) \\ \vdots \quad \vdots \quad \ddots \quad \vdots \\ k\left(\boldsymbol{x}^{(n)}, \boldsymbol{x}^{(1)}\right) k\left(\boldsymbol{x}^{(n)}, \boldsymbol{x}^{(2)}\right) \cdots k\left(\boldsymbol{x}^{(n)}, \boldsymbol{x}^{(n)}\right) \end{bmatrix}, \tag{17}$$

$$\boldsymbol{y}' = \left[y^{(1)}, y^{(2)}, \cdots, y^{(n)}\right]^T \tag{18}$$

$$\boldsymbol{\alpha} = [\alpha_1, \alpha_2, \cdots, \alpha_n]^T \tag{19}$$

The kernel matrix $\boldsymbol{K}$ is a positive definite symmetric matrix due to the definition of the kernel function.

Positive definite $\quad \boldsymbol{z}^T\boldsymbol{Kz} > 0, \quad \boldsymbol{z}$ is arbitrary $\quad$ (20)

Symmetric $\quad \boldsymbol{K}^T = \boldsymbol{K} \quad$ (21)

Regression coefficients are determined so that the residual from learning data is minimized. In this case, the extreme solution for α_i is equivalent to the minimum residual solution. Therefore, $\boldsymbol{\alpha}$ is determined as satisfying

$$\frac{\partial R_0}{\partial \boldsymbol{\alpha}} = -2\boldsymbol{K}^T(\boldsymbol{y}' - \boldsymbol{K\alpha}) = \boldsymbol{0}.$$

As a solution of this equation, we obtain

$$\boldsymbol{y}' = \boldsymbol{K\alpha}. \tag{22}$$

Hence, regression coefficients are derived as

$$\boldsymbol{\alpha} = \boldsymbol{K}^{-1}\boldsymbol{y}'. \tag{23}$$

The above $\boldsymbol{\alpha}$ obtained in this way makes the residual zero and $\hat{y}$ in Eq. 14 agrees with all learning data. However, since the series expansion of the kernel function is represented by a higher order power series, it violently vibrates at the interpolation point in the interval of the learning data, and cannot explain the phenomenon well. This is said to be over-fitting. In order to avoid over-fitting, it is sufficient to relax the degree of fitting to the learning data moderately. Moderating moderately is called generalization. In the following section the constraint condition is explained to make generalization.

3.3 REGRESSION COEFFICIENTS OBTAINED WITH GENERALIZATION

A Lagrange multiplier method is effective for obtaining the extreme (optimum) solution in a constraint condition. When a constraint condition is defined as $\|f\|^2 = c$ and unknown multiplier is denoted as λ, the Lagrange function is expressed as

$$L(\boldsymbol{\alpha}, \lambda) = R_0(\boldsymbol{\alpha}) + \lambda(\|f\|^2 - c). \tag{24}$$

As a constraint condition the orthogonality of $\boldsymbol{\alpha}$ and $\boldsymbol{y}'$is applied. If the both is orthogonal,

$$\boldsymbol{\alpha}^T\boldsymbol{y}' = 0. \tag{25}$$

is obtained. Then, since the relationship between $\boldsymbol{\alpha}$ and $\boldsymbol{y}'$is independent, the learning effect becomes small and it is relax the degree of fitting to the learning data moderately. Applying this constraint condition for the Lagrange function gives

$$L(\boldsymbol{\alpha}, \lambda) = R_0(\boldsymbol{\alpha}) + \lambda\boldsymbol{\alpha}^T\boldsymbol{y}' \tag{26}$$

and substituting this equation for Eq. 22 provides

$$L(\boldsymbol{\alpha}, \lambda) = R_0(\boldsymbol{\alpha}) + \lambda\boldsymbol{\alpha}^T\boldsymbol{K\alpha}. \tag{27}$$

The larger the value of λ is, more moderate the fitting to the learning data and the vibration at the interpolation point are relaxed. However, if λ is too large, the regression model becomes meaningless. The partial differential of Eq. 27 with respect to $\boldsymbol{\alpha}$ is

$$\frac{\partial L(\boldsymbol{\alpha}, \lambda)}{\partial \boldsymbol{\alpha}} = -2\boldsymbol{K}^T(\boldsymbol{y}' - \boldsymbol{K\alpha}) + 2\lambda\boldsymbol{K\alpha} = 0. \tag{28}$$

Hence, we obtain

$$\boldsymbol{\alpha} = (\boldsymbol{K} + \lambda\boldsymbol{I})^{-1}\boldsymbol{y}'. \tag{29}$$

The parameter λ is called the regularization parameter generalizing the regression model. $\boldsymbol{I}$ is a unit matrix. The smaller the value of λ is, the less the generalization is. If the generalization is too little, the over-fitting to the learning data occurs and the interpolation value vibrates violently. On the other hand, if the generalization is too high, the interpolation value becomes too moderate for comparing with the learning data.

3.4 REGRESSION COEFFICIENTS OBTAINED WITH CONSTRAINT CONDITION

A condition that the kernel regression model gives a fixed value y_p at $\boldsymbol{x}_p$ is

$$y_p - \boldsymbol{\alpha}^T\boldsymbol{k}(\boldsymbol{x}_p) = 0.$$

Another condition that the kernel regression model gives a gradient $\boldsymbol{z}_q$ at $\boldsymbol{x}_q$ is

$$\boldsymbol{z}_q - \left[\frac{\partial \boldsymbol{k}(\boldsymbol{x}_q)}{\partial \boldsymbol{x}_q}\right]^T \boldsymbol{\alpha} = \boldsymbol{0}.$$

Where, $\boldsymbol{k}(\boldsymbol{x}_q)$ is defined as the following vector.

$$\boldsymbol{k}(\boldsymbol{x}) = \begin{Bmatrix} k(\boldsymbol{x}, \boldsymbol{x}^{(1)}) \\ k(\boldsymbol{x}, \boldsymbol{x}^{(2)}) \\ \vdots \\ k(\boldsymbol{x}, \boldsymbol{x}^{(3)}) \end{Bmatrix}.$$

In order to add these conditions to the Lagrange function as a constraint condition, we define the square of residual as follows and consider the limitation.

$$r_1^{(i)}(\boldsymbol{\alpha}) = \left\{y_p^{(i)} - \boldsymbol{\alpha}^T \boldsymbol{k}(\boldsymbol{x}_p^{(i)})\right\}^2 \to 0.$$

$$r_2^{(j)}(\boldsymbol{\alpha}) = \left\| \boldsymbol{z}_q - \left[\frac{\partial \boldsymbol{k}(\boldsymbol{x}_q)}{\partial \boldsymbol{x}_q}\right]^T \boldsymbol{\alpha} \right\|^2 \to 0.$$

Then, the Lagrange function satisfying several constraint conditions is expressed as

$$L(\boldsymbol{\alpha}, \lambda) = R_0(\boldsymbol{\alpha}) + \lambda \boldsymbol{\alpha}^T \boldsymbol{K} \boldsymbol{\alpha} + \sum_{i=1}^{n_p} \lambda_1^{(i)} r_1^{(i)}(\boldsymbol{\alpha}) + \sum_{j=1}^{n_q} \lambda_2^{(j)} r_2^{(j)}(\boldsymbol{\alpha}) \tag{30}$$

Minimizing the Lagrange function gives the following equation.

$$\boldsymbol{\alpha} = \boldsymbol{B}^{-1} \boldsymbol{u} \tag{31}$$

$$\boldsymbol{u} = \boldsymbol{y} + \boldsymbol{K}^{-1} \left\{ \sum_{i=1}^{n_p} \lambda_1^{(i)} y_1^{(i)} \boldsymbol{k} + \sum_{j=1}^{n_q} \lambda_2^{(j)} \left[\frac{\partial \boldsymbol{k}}{\partial \boldsymbol{x}_q}\right] z_q^{(j)} \right\} \tag{32}$$

$$\boldsymbol{B} = \boldsymbol{K} + \lambda \boldsymbol{I} + \boldsymbol{K}^{-1} \left\{ \sum_{i=1}^{n_p} \lambda_1^{(i)} \boldsymbol{k}\boldsymbol{k}^T + \sum_{j=1}^{n_q} \lambda_2^{(j)} \frac{\partial \boldsymbol{k}}{\partial \boldsymbol{x}_q} \left[\frac{\partial \boldsymbol{k}}{\partial \boldsymbol{x}_q}\right]^T \right\} \tag{33}$$

3.5 CROSS VALIDATION

Cross validation is used to confirm the appropriateness of model parameters and regularization parameters. Cross validation (CV) is a method of evaluating the performance of the estimated statistical model based on sample data. In general, sample data is divided into learning data and test data, and the error between statistical model estimated from learning data and test data is obtained. A combination of learning data and test data is prepared by a plurality of patterns, and the parameters of the statistical model are tuned so that the average of the error is minimized. In the case of the kernel regression model, Leave-one-out cross validation (one piece withdrawal CV) using one test data only can be applied. By using the following formula, CV error can be calculated efficiently without separating learning data and test data.

$$\mathrm{CV} = \frac{1}{n} \sum_{i=1}^{n} \left(\frac{y^{(i)} - \hat{y}^{(i)}}{1 - H_{ii}} \right)^2. \tag{34}$$

Where,

$$\hat{\boldsymbol{y}}^{(i)} = \boldsymbol{K} \boldsymbol{\alpha},$$
$$\boldsymbol{H} = (\boldsymbol{K} + \lambda \boldsymbol{I})^{-1} \boldsymbol{K}.$$

4. SHAFT POWER ANALYSIS

4.1 OUTLINE OF VOYAGE DATA

We study a practical statistical model using simulated on-board monitoring data of KCS container ship. Table 1 summarizes the principal particulars of the ship.

Since the KCS container ship (KCS) is virtual, the voyage simulation data is assumed to be the simulated on-board monitoring data. Since the ship performance can be perfectly calculated, it is useful for the basic verification of statistical model. Data were taken as the hourly encountered sea-state and various short-term ship responses obtained in a one-year voyage on the North-Pacific route between Japan and North America. It takes about seven days for loading cargo on/off in each country and about 8 days for sailing in both outbound and return. Therefore, 30 days per voyage is required, it will be 12 voyages (round-trip) in one year. The number of data is about 5500 per a ship response.

The outline of the computer simulation is as follows. The engine characteristics were fixed in a short-term sea-state. Based on this characteristic, two kinds of the equilibrium equation are provided. One is the equilibrium of the short-term steady hydrodynamics force and the propeller thrust acting on the hull in the encountered sea-state, and the other is the equilibrium of the engine torque and the propeller torque. Solving these equations, we obtain the ship speed and the engine motion. The tracking path is obtained by multiplying the ship speed and the step time. This calculation is repeated from the departure port to the arrival port. Wave steady-state force among resistance components (Longitudinal force, lateral force, turning force) are made into a database for each ship speed, wave period, wave direction, and the steady-state force at an

Table 1 Principal particulars of KCS.

Item	Symbol	Unit	KCS
Type			container
Service speed	V	knots	25.0
Length of overall	L_{oa}	m	238.3
Length between p.p.	L_{pp}	m	230.0
Breadth	B	m	32.2
Depth	D	m	19.0
Draft(Midship)	d_m	m	10.8
Displacement	∇	m^3	52054
Route			JP-USA
Period	T	days	365

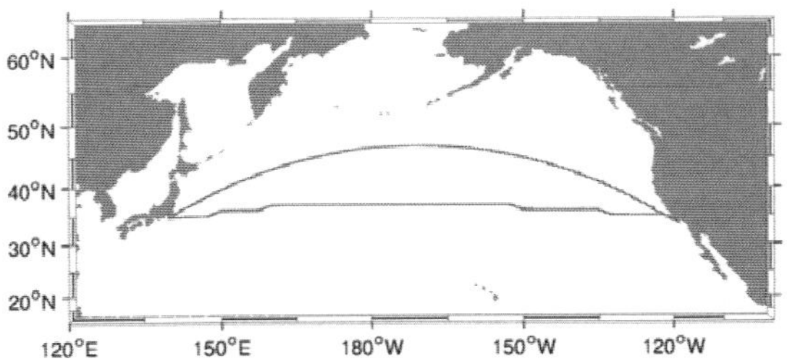

Fig. 1 Voyage route of KCS container.

arbitrary ship speed is calculated by interpolation calculation. This database is provided by the RIOS system (http://www.rios.eng.osaka-u.ac.jp/) based on EUT (Enhanced Unified Theory) developed by Kashiwagi. The wave spectrum is the ITTC spectrum. Wind pressure is estimated by Fujiwara. The propulsive resistance is calculated from the frictional resistance equation of the equivalent flat plate of Schönherr and improved formula by Yoshimura. Propeller thrust and torque are obtained from the open chart of quadratic curve approximation of advance constant.

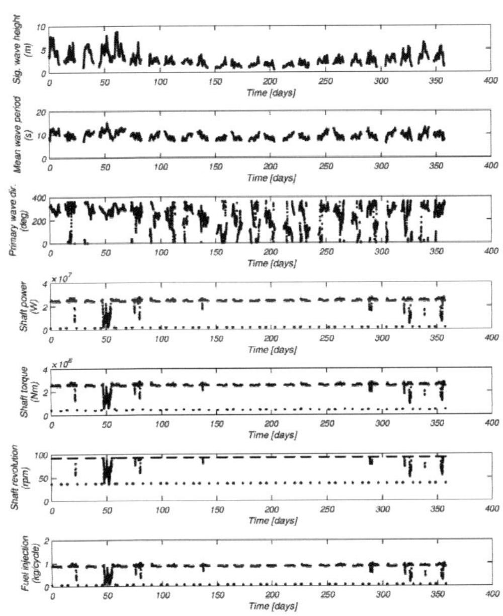

Fig. 2 Short-term ship responses of KCS.

4.2 DATA SCRUTINY

The raw data of the on-board monitoring data includes data during berthing and maneuvering in the port, data in a system shutdown due to a power interruption, abnormal data due to some trouble. These data are meaningless for estimating ship performance, and analysing including these data may lead to erroneous results, so it must be deleted from the data set. This is called data scrutiny. In the voyage simulation, such data does not exist in principle. The route is shown in Fig. 1 and an example of monitoring data as a short-term ship response is shown in Fig. 2

4.3 NORMALIZATION OF DATA

Since monitoring data has dimensions (unit), there is an uneven distribution in magnitude of data fluctuation. If this uneven distribution is remarkable, the analysis result may be dominated by data with large fluctuations. Normalization of data is pre-processing to eliminate the deviation of data fluctuation due to such dimensions and enable accurate analysis. Here, independence is shown as normalization.

Let an input vector variable define as follows:

$$\boldsymbol{\xi} = \left[V_{rcx}, V_{rax}, V_{ray}, H_{1/3}, G_{rx}, G_{ry}, X_{r4}, X_{r5}, \beta, D_m\right]^T \quad (35)$$

and independence formulate as follow:

$$\boldsymbol{x} = \boldsymbol{WDVC\xi} \quad (36)$$

$\boldsymbol{C}$ is a diagonal matrix with the inverse of standard deviation of $\boldsymbol{\xi}$. The variance of component of $\boldsymbol{C\xi}$ is naturally unified to one. $\boldsymbol{V}$ is a matrix which makes the covariance matrix of $\boldsymbol{\xi}$ diagonal. Diagonalization is performed by dispersion maximization according to the method of principal component analysis. Then, vector components of $\boldsymbol{VC\xi}$ are uncorrelated each other. Further, normalization of variance is carried out by $\boldsymbol{D}$ which is a diagonal matrix with the inverse of standard deviation of $\boldsymbol{VC\xi}$.

$$\boldsymbol{VC}\mathrm{Cov}[\boldsymbol{\xi}, \boldsymbol{\xi}]\boldsymbol{C}^T\boldsymbol{V}^T \rightarrow \text{Diagonal matrix}$$
$$\text{Component of } \boldsymbol{VC\xi} \rightarrow \text{Maximum of varinace}$$

$\boldsymbol{W}$ is an orthogonal matrix that separates the vector components of $\boldsymbol{DVC\xi}$. Since it is an orthogonal matrix, the covariance matrix of the original component remains diagonalized. Independence is also uncorrelated, so there is no contradiction. $\boldsymbol{W}$ is determined to maximize (or minimize) fourth order cumulants, assuming that the definition of independence is the maximum divergence from the normal distribution of the components. Since $\boldsymbol{W}$ is an orthogonal matrix, the variance of the vector components of $\boldsymbol{WDVC\xi}$ remains at one.

$$\boldsymbol{WDVC}\mathrm{Cov}[\boldsymbol{\xi}, \boldsymbol{\xi}]\boldsymbol{C}^T\boldsymbol{V}^T\boldsymbol{D}^T\boldsymbol{W}^T$$
$$\rightarrow \text{Diagonal matrix}$$
$$\text{Component of } \boldsymbol{WDVC\xi}$$
$$\rightarrow \text{Maximum/Minimum of 4th order cumulant}$$

The relationship between $\boldsymbol{x}$ and $\hat{P}_s$ is expressed using the kernel regression model as follows.

$$\hat{P}_s(\boldsymbol{x}) = \sum_{j=1}^{n} \alpha_j k\left(\boldsymbol{x}, \boldsymbol{x}^{(j)}\right), \quad (37)$$

Table 2 Sea-state and primary resistance expected by zero monitored data.

Case	Expected sea-state	Monitored data for constraint condition	Performance
A.1	No wind-wave, No swell, No wind (Relative head wind)	Concurrent P_s, V_{rax}, $H_{1/3}$, G_{rx}, G_{ry} with maximum V_{rcx} in $H_{1/3} \sim 0$, $X_{r4} \sim 0$, $X_{r5} \sim 0$, $V_{rcx} \sim V_{rax}$ for each D_m (Constraint on $[V_{ray}, X_{r4}, X_{r5}, \beta] = \mathbf{0}$)	Max. shaft power in calm sea (with relative head wind by crusing)
A.2		Concurrent P_s, V_{rax}, $H_{1/3}$, G_{rx}, G_{ry} with maximum V_{rcx} in $H_{1/3} \sim 0$, $V_{rcx} \sim V_{rax}$ for each D_m (Constraint on $[V_{ray}, X_{r4}, X_{r5}, \beta] = \mathbf{0}$)	
A.3		Concurrent P_s, V_{rax}, $H_{1/3}$, G_{rx}, G_{ry} with maximum V_{rcx} in $X_{r4} \sim 0$, $X_{r5} \sim 0$, $V_{rcx} \sim V_{rax}$ for each D_m (Constraint on $[V_{ray}, X_{r4}, X_{r5}, \beta] = \mathbf{0}$)	
B	Following wind-wave, No swell, No relative wind (Tail wind)	Concurrent P_s, $H_{1/3}$, G_{rx}, G_{ry} with maximum V_{rcx} in $X_{r4} \sim 0$, $X_{r5} \sim 0$, $V_{rax} \sim 0$, $G_{ry} \sim 0$ for each D_m (Constraint on $[V_{rax}, V_{ray}, X_{r4}, X_{r5}, \beta] = \mathbf{0}$)	Max. shaft power in following wave without relative wind and ship motion
C	Wind-wave, No swell, Head and tail wind	Concurrent P_s, V_{rcx}, $H_{1/3}$, G_{rx}, G_{ry} with maximum V_{rax} in $X_{r4} \sim 0$, $X_{r5} \sim 0$, $V_{ray} \sim 0$ for each D_m (Constraint on $[V_{ray}, X_{r4}, X_{r5}, \beta] = \mathbf{0}$)	Max. shaft power without relative beam wind and ship motion

4.4 BOUNDARY CONDITION OF SHAFT POWER FUNCTION

Power curve in calm sea condition is the lower limit of propulsive performance. Accurate estimation of the lower bound improves the reliability of the statistical model. This lower limit is a zero power at zero ship speed, and a cruising power near the design speed. The interval between both powers can be connected smoothly like a cubic curve. When the statistical model is identified from on-board monitoring data, it is not always possible to explain these zero and cruising powers accurately. Therefore, we consider adding these matters as a constraint condition.

It is naturally obvious that the shaft power is zero when the relative ship speed for water is zero. Cruising power is generally presumed from experiments and sea trials, etc. However, it is very difficult to exhaustively cover the operational conditions, so we use on-board monitoring data. In this case, it is not necessary that the power is exactly the cruising power, and if the relationship between the power and the ship speed etc. is accurate, data in the vicinity of the cruising power may be used. Although there are many data in the vicinity of the cruising power in waves, it is rare in calm sea.

When there is no disturbance and the relative ship speed for water is zero, the shaft power should be zero, and the power gradient for various ship responses should be also zero. Then, the conditions are as follows.

$$y_p^{(1)} = 0,$$
$$\boldsymbol{\xi}_p^{(1)} = \mathbf{0} \rightarrow \boldsymbol{x}_p^{(1)} = \mathbf{0},$$

$$\boldsymbol{z}_q^{(1)} = \mathbf{0},$$
$$\boldsymbol{\xi}_q^{(1)} = \mathbf{0} \rightarrow \boldsymbol{x}_q^{(1)} = \mathbf{0},$$

Since strong constraints are needed here, $\lambda_1^{(1)} = \lambda_2^{(1)} = 1000$ is assumed.

Table 2 shows how to distinguish power near-cruising operation in a calm sea condition. A data set whose ship motion, wave height, etc. are near zero at every draft and the relative ship speed for water is the largest is basically selected as a constraint condition. At this time, ship motion, wave height, etc. shall be zero. However, since it is not exactly zero, we have a weak constraint as $\lambda_1^{(1)} = 1$. Constraint conditions include not only in perfect calm sea condition, but also in the vicinity of calm sea condition. A.1 to 3 are methods of selecting calm sea condition data, B is a method of selecting data with no relative wind and no ship motion in a following sea condition, and C is a method of selecting data with no crosswind and no ship motion in a all directional wave condition. In this paper, the constraint condition data was selected by the method of A.1, B, and C.

4.5 ESTIMATION OF SHAFT POWER FUNCTION

Using the kernel regression model, the component of the shaft power was estimated. The parameters c, p of the kernel function are as follows.

$$c = 1, p = 9$$

p is a natural number that minimizes CV value.

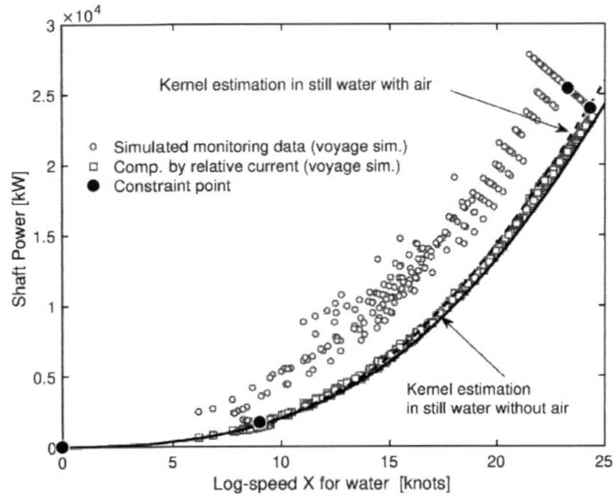

Fig. 3 Total shaft power and its component in still water.

Fig. 3 shows the shaft power component in calm sea condition due to the still water resistance. The solid line is the component excluding the wind resistance of the self-running. It was estimated by

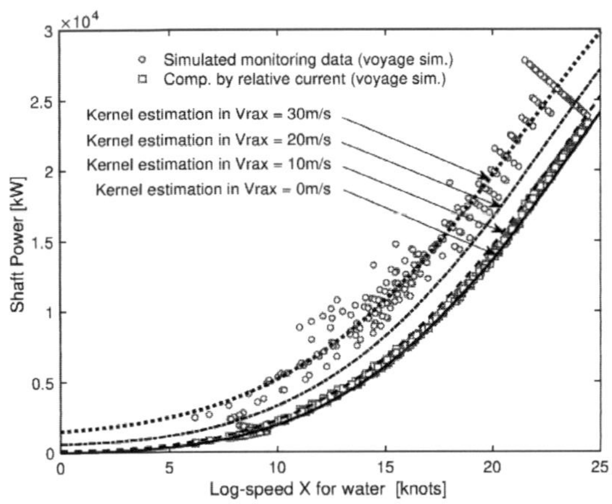

Fig. 4 Total shaft power in winds.

$$\xi = [V_{rcx}, 0, \cdots, 0, D_m]^T \tag{38}$$

On the other hands, the broken line indicates the component including the wind resistance of the self-running. It was estimated by

$$\xi = [V_{rcx}, V_{rcx}, \cdots, 0, D_m]^T \tag{39}$$

The black circle is the constraint point defined by the constraint condition. Although it restrains at four points, at the zero point, the power gradient is also constrained. It is a strong constraint at the zero point and a weak constraint elsewhere. By imposing such constraint conditions, the reliable analysis becomes possible. White circles and white squares are shaft powers as simulated on- board monitoring data. The white circle is the shaft power obtained by summing all components and is the amount measured on an actual ship. White squares are components in calm sea condition excluding the wind resistance. The white squares and the solid lines are

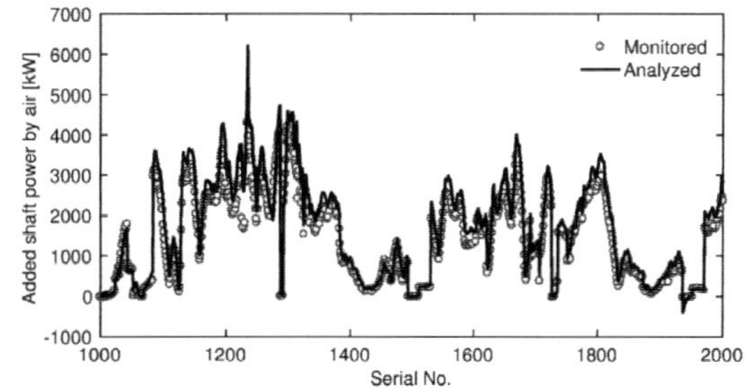

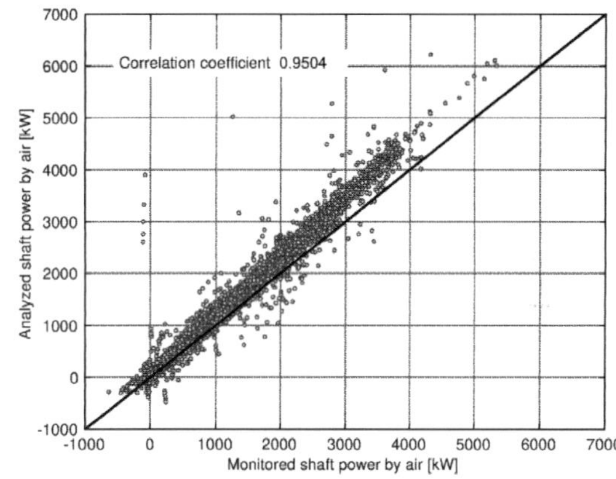

Fig. 6 Comparison between monitor and estimation of added shaft power in winds.

almost in agreement, which shows that they are correctly estimated. Although the influence of the wind resistance of self-running is small, it is important to evaluate the wind resistance of the hull.

Fig. 4 shows the estimation result of the relation between the magnitude of the wind speed and the power increase. We compare winds with relative wind speeds of 0, 10, 20 and 30 m/s. Since the power increases according to the magnitude of the wind speed, a reasonable estimation is made. In addition, it is shown that shaft power is necessary so that even if the relative ship speed for water is zero, it would not be drifted by the wind.

Fig. 5 shows the added shaft power due to wind resistance during voyage, and the correlation between monitoring data and estimated value is shown in Fig. 6. This was estimated by

$$\Delta P_{s(wind)} = P_s - \hat{P}_s(\xi) \tag{40}$$

$$\xi = \left[V_{rcx}, 0, 0, H_{1/3}, G_{rx}, G_{ry}, X_{r4}, X_{r5}, \beta, D_m\right]^T \tag{41}$$

Although the monitoring data and the estimated value show a high correlation (correlation coefficient 0.9504), the estimated value tends to increase in the region where t he shaft power is large.

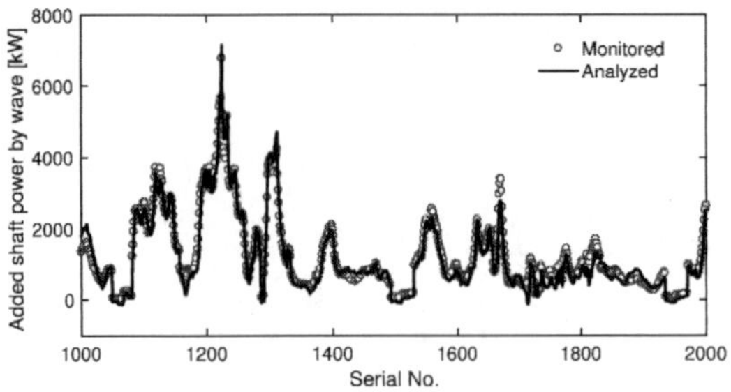

Fig. 7 Added shaft power in waves.

Fig. 7 shows the added shaft power in waves due to the wave resistance during the voyage, and the correlation between the monitoring data and the estimated value is shown in Fig. 8. This was estimated by

$$\Delta P_{s(wave)} = P_s - \hat{P}_s(\xi) \tag{42}$$

$$\xi = \left[V_{rcx}, V_{rax}, V_{ray}, 0, \cdots, 0, \beta, D_m\right]^T \tag{43}$$

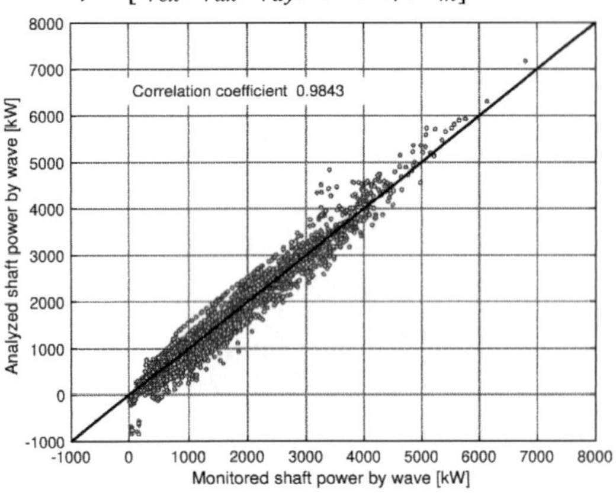

Fig. 8 Comparison between monitor and estimation of added shaft power in waves.

The monitoring data and the estimated value show a high correlation (correlation coefficient 0.9843). Although it has large variance, it can be estimated well on average.

5. CONCLUSIONS

We proposed a method for estimating the shaft power function of a ship by a statistical approach using on-board monitoring data. By expressing the shaft power function by the kernel multivariate regression model and by independentizing the data for the model identification, it is possible to analyse the data in a dimensionless manner. We showed that it is possible to estimate the shaft power function accurately by adding the performance boundary condition as the constraint condition. By using the shaft power function, it was shown that the added power due to disturbance such as winds and waves can be accurately estimated.

For the future works, we will use the on-board monitoring data on the actual ship to improve the shaft power function and apply it to the service performance analysis.

6. ACKNOELEDGEMENTS

This work was supported by JSPS KAKENHI Grant Number JP15H04213.

7. REFERENCES

1. MINOURA, M., NAITO, S, 'Statistical Estimation of Ship Speed Loss by Monitoring Data' *Proc. Of 6th Osaka Colloquium on Seakeeping and Stability of Ships (OC2008)* pp.37-42, 2008.

2. MINOURA, M. and KOIKE, K.: 'Multivariate Short-term Prediction of Ship Response Using On-board Measurement Data', *Proc. of 11th International Marine Design Conference*, Vol. 2, pp. 49-61, 2012.

3. HYVARINEN, A., OJA E. and KARTHUNEN J., 'Independent Component Analysis', *John Wiley & Sons, Inc.*, 2001.

4. MMINOURA, M., OSADA, S., HASE, N. and NAITO S., 'Oceangoing Simulation with Deliberate Speed Loss and Course Change', *Proc. of Advanced Maritime Engineering Conference 2008 (AMEC2008)*, pp.423-430, 2008.

5. MINOURA, M.inoura, NAITO S., 'Stochastic Sea Climate Simulation Based on Hindcast Data', Proc. of 16th International Offshore and Polar Engineering Conference, Vol. 3, pp94-101, 2006.

6. KASHIWAGI, M., MIZOKAMI, S., YASUKAWA, H. and FUKUSHIMA, Y., 'Prediction of Wave Pressure and Loads on Actual Ships by the Enhanced Unified Theory',*Proc. of 23rd International Symposium on Naval Hydrodynamics,* pp.95-109, 2000.

8. AUTHORS BIOGRAPHY

Munehiko Minoura holds the current position of associate professor at Osaka University. He is responsible for ship performance analysis using the statistical method, in particular seakeeping performance. In this present, he is struggling the performance analysis using on-board monitoring data. He worked for Mitsui Engineering and Shipbuilding Company in Japan. In that time, he had experience of the fatigue analysis based on FEM.

SHIP MACHINERY FUZZY CONDITION BASED MAINTENANCE

M. Cheliotis and **I. Lazakis**, Department of Naval Architecture, Ocean and Marine Engineering, University of Strathclyde, UK

SUMMARY

One of the main factors that influence the performance of the global fleet is the physical condition of its vessels. Well maintained ships exhibit higher reliability, profitability and safety. Even though machinery failures are inevitable, their occurrences can be minimised, or even predicted, when predictive maintenance schemes are implemented. This paper examines how a condition based maintenance approach can be used in conjunction with modern approaches from data analytics. The methodology combines the use of a quantitative Dynamic Fault Tree Analysis (DFTA) with data clustering to identify critical ship systems. Data from critical machinery operating conditions are handled and using Fuzzy Logic (FL) risk indices are obtained, based on which maintenance actions are suggested. The use of expert judgment, Original Equipment Manufacturer (OEM) thresholds and sensorial data form the input for presented methodology.

NOMENCLATURE

AI	Artificial Intelligence
I^b	Birnbaum Importance Measure
ConMon	Condition Monitoring
CM	Corrective Maintenance
I^{cr}	Criticality Importance Measure
EMSA	European Maritime Safety Agency
DFT	Dynamic Fault Tree
DFTA	Dynamic Fault Tree Analysis
FES	Fuzzy Expert System
FL	Fuzzy Logic
FMEA	Failure Mode Event Analysis
FO	Fuel Oil
IACS	International Association of Classification Societies
IM	Importance Measure
LO	Lubricating Oil
ME	Main Engine
MC	Markov Chains
OEM	Original Equipment Manufacturer
PM	Predictive Maintenance
PMS	Planned Maintenance Scheme

1. INTRODUCTION

The importance of shipping in the global economy is unquestionable. It is also clear, that the condition of the world's merchant fleet and the ship maintenance is, and always has been a pressing issue for the maritime industry. Ship failures can disrupt supply chains, undermine human safety, endanger the environment and result in expensive corrective actions [1][2]. For these reasons, maritime maintenance is a heavily regulated issue on both national and international levels.

In general, breakdowns can be dealt with either in a corrective, preventive or predictive manner. Corrective Maintenance (CM) is the oldest maintenance approach, whereas predictive the most recent. CM is a reactive approach and initiates post failure. In the modern maritime industry, CM should only be used as a starting point, or for non-critical equipment [3]. On the other hand, the minimization of ship failures is the area of interest for preventive maintenance [4], whereas their prediction for Predictive Maintenance (PM) and Condition Monitoring (ConMon) [5].

Currently, all the maintenance approaches (corrective, preventive and predictive) are being used, but in different extents. Interestingly, the past decades have seen the development of different tools used mainly in PM and ConMon, ranging from vibration monitoring and acoustic analysis to thermography and lubricating oil analysis [6]. Most of these tools aim to collect sensorial data from various ship equipment. These tools are based on algorithms and processes that translate the recorded data into reliability indices.

Nowadays, predictive maintenance is refocusing on techniques and methodologies from the areas of data analytics, machine learning and Artificial Intelligence (AI). The aim is to make direct or indirect reliability forecasts. In the context of this paper, indirect forecasts focus on the prediction of physical parameters based on data analytics and machine learning [7]. The forecasted parameters are then interpreted to risk and reliability indices based on risk assessment methodologies.

Nonetheless, ConMon and PM must be addressed with caution. Since many of their tools are heavily dependent on data, the creation of large and cumbersome databases is a possibility. Even though big data are becoming ever more present in the maritime industry, their use must originate from their necessity. Organisations implementing ConMon and PM must have a very clear understanding of the required data for successful maintenance. Key questions that must be answered include:

- Which systems should be monitored?

- Which components should be monitored?
- Which parameters should be assessed?

By successfully answering these questions we can make the transition from big data to smart data.

The present paper follows the previously described trend. The methodology initiates with the selection of the ship critical equipment (Section __). By using DFTA, the criticality of ship systems is assessed and by using data clustering (k-means) the critical equipment for further analysis is selected. Then, measurable parameters are selected for each of the critical equipment through the use of statistical tools (Section __). Lastly, FL and Fuzzy Expert Systems (FES) are used to translate raw data into risk indices (Section __).

2. LITERATURE BACKGROUND

It is true that poor maintenance practices can undermine the safety of ships which in turn increases the risk of accidents. Such accidents can cause environmental and economic damage, as well as loss of life. Lack of proper maintenance can be a major accident contributor, especially if it is combined with design flaws and human errors. The absence of apt maintenance can manifest its self in the form of engine trouble, propulsion trouble, mechanical issues, generators malfunctions, electrical problems and plumbing failures. In detail, according to the European Maritime Safety Agency (EMSA), 12591 casualties and incidents were reported between 2011 and 2015. From these, 27% were caused by poor maintenance resulting in equipment failure. The following table summarizes the contribution of maintenance as an underlying cause of the accidents reported between 2011 and 2015 as a function of ship type [8].

Table 1: Maintenance accident contribution per ship type

Ship Type	***Maintenance Accident Contribution***
Cargo	*20%*
Fishing	*17%*
Passenger	*28%*
Service	*21%*

2.1 MAINTENANCE REGULATORY BACKGROUND

The importance of maintenance becomes apparent in all the industries and especially in the maritime sector. Because of its importance, ship maintenance is a heavily regulated issue. Ship maintenance is concerned with the performance and condition of the shipboard machinery, as well as with the hull and its appendages.

To tackle the issue of safety in a complete way, the International Maritime Organisation adopted the International Management Code for the Safe Operation of Ships and for Pollution Prevention (ISM Code) in 1993. The ISM Code is divided into 13 sections, each of which covers a different aspect of ship management and operation. A particular section of interest is section 10, which is concerned with the maintenance of the ship and its equipment. In detail, it is specified that it is the responsibility of the shipping company to make sure that their assets are maintained according to relevant rules and regulations. The maintenance should be carried out in a structured and planned way by appropriate personnel, whilst having safety as a priority [9].

Apart from IMO's regulations, there are other governing and industry bodies that regulate the issue maritime of maintenance. Classification societies (ABS, BV, LR, etc.) and the International Association of Classification Societies (IACS) are organizations which have regulations concerning maintenance. The purpose of the classification societies and consequently IACS is to confirm the reliability of ship systems and to verify that systems and components are maintained appropriately. Therefore, such organizations develop their own rules and standards. However, classification societies and IACS are not guarantors of safety, as they do not have control over how a ship is manned and operated.

2.2 PREDICTIVE MAINTENANCE

Under the regulatory requirements described above, a shipping company must adopt an appropriate and adequate maintenance strategy. This strategy includes a plethora of different approaches. It can include corrective actions for non-critical equipment, and for equipment with an easy access to spare parts. Preventive maintenance can also be involved. It can form a time-based upkeeping routine embedded in the company's Planned Maintenance Scheme (PMS).

The present paper will focus on the newest maintenance approach, PM. In PM, also known as CBM, the machinery and the ship systems are monitored constantly and their condition is assessed to schedule their maintenance accordingly. The underlying principle of CBM is that maintenance is a function of the equipment's need and state [10]. What sets CBM apart from other maintenance schemes is that it tries to calculate the machineries' actual remaining useful life, rather than relying on industrial average statistics [11]. PM is ideal when trying to address the issues of over-maintaining and under-maintaining the equipment. Videlicet, predictive maintenance is used to address some of the shortcomings of preventive and CM practices. CBM is a more modern and dynamic approach to maritime maintenance and asset management. The following table summarises the benefits and drawbacks of PM, as well as the tools it uses.

Table 2: Predictive maintenance tools and characteristics

Benefits	***Drawbacks***	***Tools***
Lower maintenance cost	*Increased capital investment*	*Thermography*
Reduced breakdowns	*Need for experienced personnel*	*Vibration Monitoring*
Increased availability	*Benefits appear in the long-run*	*Acoustic Analysis*
Better maintenance plan	*Additional resources for data handling*	*Current Analysis*
Increased equipment utilization efficiency	*Heavily dependable on sensor output*	*Lubricating Oil Analysis*

2.3 RELATED STANDARDS

Condition Based Maintenance and Predictive Maintenance techniques are subject to international standards of quality. The most important standards are set by British Standards (BS) and International Organization for Standardization (ISO). The following list contains some of the most widely used standards [12]:

- BS/ISO 13613 (2011) deals with the overall importance of machinery maintenance.
- BS/ISO 17905 (2015) focuses on the maintenance of the machinery of container ships.
- BS/ISO 7919 (2009) examines vibration measurements for rotating shafts.
- BS/ISO 13381 (2015) standardises prognostic techniques for CBM
- BS/ISO 17359 (2011) examines multi-parameter diagnostic techniques.
- BS/ISO 13379 (2012) concerns data interpretation for the use of marine CBM.

3. METHODOLOGY

The proposed methodology is based on monitoring key equipment's operating parameters to develop risk indices. Also, this methodology provides a comprehensive approach to the selection and identification of the critical equipment.

3.1 METHODOLOGY OVERVIEW

The methodology initiates with the identification of the ship's critical components. The ship's systems are depicted in a Dynamic Fault Tree (DFT) structure. The DFTA calculates the criticality Importance Measures (IMs), which are then plotted together with additional criteria. In the plot, a data clustering algorithm is implemented to organize the data. From the resulting clusters, components are selected based on specific criteria.

After examining the selected equipment, the measurable parameters for the analysis are selected. At this stage, statistical tools are used to identify "connections" between variables.

Lastly, the selected parameters are assessed by using a FES which is a tool originating from the field of AI. The following figure illustrates the described methodology.

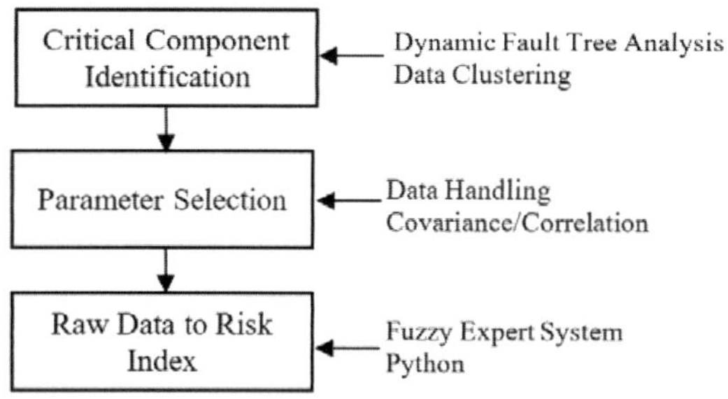

Figure 1: Methodology overview

3.2 CRITICAL EQUIPMENT SELECTION

One of the tools used in the selection of the critical equipment is the DFTA. Fault trees are a proven and established tools used in many industries for various reliability, safety and risk calculations [13].

FTA is a top-down approach that examines the system's internal arrangement and the effect of both internal and external disturbances. It can be used for both qualitative and quantitative purposes. FTA functions based on classical Boolean operators and DFTA operates based on Markov Chains (MC). FTA is a static assessment tool, whereas DFTA is a dynamic one [14]. A sample from the resulting DFT structure is shown in the following figure.

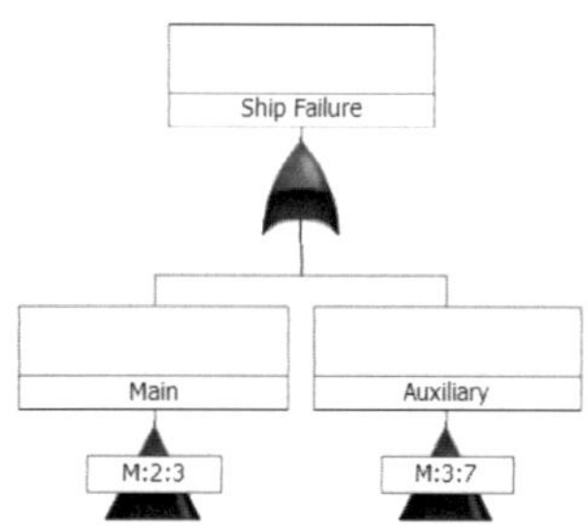

Figure 2: DFTA representation

For the quantitative DFTA, failure rates were used as inputs and the appropriate calculation method was selected. Moreover, the critical ty IM (I^{cr}) and the Birnbaum IM (I^{b}) were calculated based on the following equations

$$I_i^{cr}(A) = P\{X/A\} - P\{X/{\sim}A\} * \frac{P\{A\}}{P\{X\}}$$

$$I_i^{B}(A) = P\{X/A\} - P\{X/{\sim}A\}$$

The obtained I^{cr} and I^{b} together with the repair costs for each of the considered components were put through a data mining process. The I^{cr} IM and I^{b} IM were used for the following reasons [15]:

- Both are well-established reliability metric
- The I^{cr} considers the components reliability. In that way, less reliable components are given more attention.
- The I^{b} does not reflect the components reliability, instead, it examines the effect of the other components' reliability.

Data mining includes several processes that aim to extract information from datasets. Data mining techniques are currently used in medical research, financial and investing institutions, AI development, computer science and lately they have been gaining more popularity in the marine sector and more specifically in ConMon and PM [16]. In general, data mining process aim to increase our understanding of collected datasets by either making predictions on variables, or by carrying out descriptive tasks (pattern recognition) [17]. The two main processes of data mining are data clustering and data classification.

Data classification is a predictive supervised machine learning approach. Supervised approaches need a training before any knowledge extraction can place for data sets. Training involves the use of data with previously known properties and corresponding states [18]. Normally, data classification is used to perform forecasts.

Data clustering is a cornerstone of data mining and is used for unsupervised machine learning [19]. It is a descriptive approach that aims to uncover set of distinct categories within the dataset under consideration. That way we can uncover groups of data with similar behavior that may have been previously unknown [20]. Therefore, in the context of this paper data clustering was used to uncover hidden information.

Regarding the clustering, the k-means algorithm was used from the available methods. The k-means aims by minimising the distance between every data point and its centroid. The objective function that is used is the following:

$$J = \sum_{i=1}^{k}\sum_{j=1}^{n}(\|x_i - v_j\|)^2$$

3.3 PARAMETER SELECTION

The reliability assessment of the selected system was based on the evaluation of specific equipment variables (pressure, temperature, etc.). The selected variables should be such that can describe the operating condition of the machinery as accurately as possible. The approach that was followed utilized two types of parameters. The first type of parameter included fault indicative variables (e.g. temperature for overheating). The second type of parameter was general performance variables (e.g. power output).

For the fault indicative variables, FMEA reports were assessed to correlate failure modes with possible measurable variables. FMEA is a bottom-up failure analysis tool, that analyses failures with their underlying causes and resulting consequences [21]. Under the same scope, manufacturer's guides and manuals were assessed. By using both FMEA and the manufacturer's documentation expert judgment and past experience are taken into account.

Similarly, for the general performance variables, data from electronic data collection platforms were used. That also allowed the exploration of hidden relationships between variables from different systems, using the covariance and correlation matrix. These statistical tools, allow us to examine whether variables vary together in a systematic way [22].

3.4 RAW DATA CONVERSION

Following the parameter selection, the chosen variables were put through a raw data conversion tool. The input of this tool was samples from the measurable variables and the output was a calculated risk index for the machinery in question.

The developed tool is presented schematically in the following figure. Its working principle is based on the use of Fuzzy Expert Systems. For each of the selected variable, an appropriate membership function is allocated. However, in all the cases a risk membership function, which is the desired output, is developed too. The function of the FES is to compare distances between each of the inputs with certain baseline values. The baseline values are obtained from the OEMs and represent operational thresholds. The output is decided based on the distances and certain rules, that are set by the designer of the FES. These rules represent operational conditions and limitations.

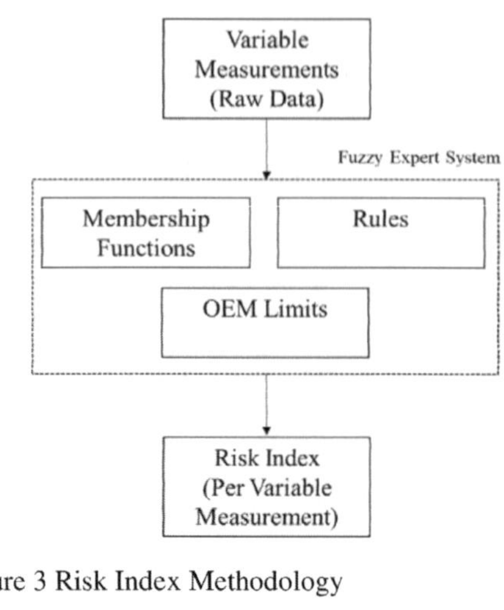

Figure 3 Risk Index Methodology

FESs are based on fuzzy logic and they are designed to extract conceptual knowledge from datasets. Fuzzy logic is designed to mathematically model vagueness and uncertainty [23]. Fuzzy logic is used to mimic human reasoning in mathematical and engineering problems. In fuzzy logic, elements can belong, with different degrees of memberships, to two different sets at the same time, which goes beyond traditional statistics. This enables us to use "degrees of truth" rather than the Boolean crisp approach [24].

For the Boolean crisp sets, an element x in the universe X has a binary degree of membership of some crisp set A. This can be mathematically represented as,

$$x_A(x) = \begin{cases} 1, & x \in A \\ 0, & x \notin A \end{cases}$$

where $x_A(x)$ represents the binary condition of membership of the element x in the set A. On the other hand, the partial membership of element x in a fuzzy set A' is described by [25].

$$\mu_{A'}(x) =\in [0,1]$$

4. CASE STUDY

The methodology described in the previous chapter is applied in a case study outlined in this section.

The process detailed in chapter 3.2, regarding the identification of critical components, is applied in a typical Panamax bulk carrier. Should be specified that provisions are in place to modify the methodology for application in other, more complex, ship types. For the FT structure, ship and engineering drawings were used. As numerical inputs, data from a PMS and expert judgment were used. Lastly, regarding the cost information used in the clustering part, mock data were produced.

The approach detailed in chapter 3.4, regarding the raw data conversion, is tested in the thrust bearing of a marine main engine. For the purpose of the analysis temperature measurements were used. The measurements were obtained hourly for a period of one month.

5. RESULTS

This section presents the results of the methodology applied to the case study described above.

5.1 CRITICAL COMPONENT SELECTION

The critical component selection is based on DFTA and data clustering. The DFTA is carried out using the PTC Windchill software, whereas the data clustering is done in python.

The selection of the critical equipment initiates with the DFTA. A sample of the resulting structure is presented in figure 2. By utilizing information from a PMS and by using the exact calculation method, reliability metrics are obtained.

These metrics (I^{cr} IM and I^{b} IM) along with cost information (mock data) were used in the next step. After being normalized the data were clustered using the k-means algorithm. From the resulting clusters, the components that belonged to the cluster whose centroid had the biggest distance from the axes origin were selected. Figure 4 shows the resulting clusters.

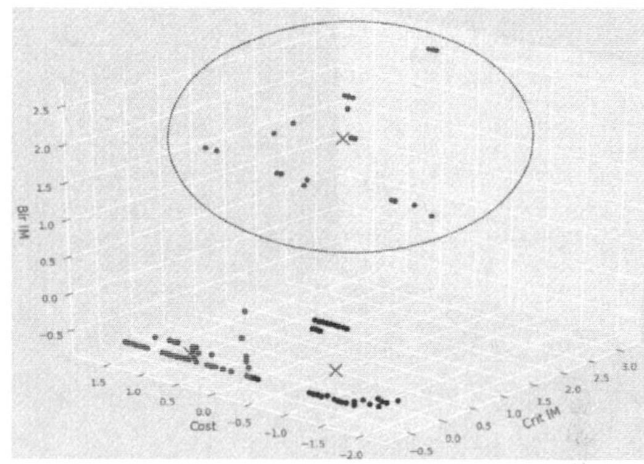

Figure 4: Cluster equipment selection

The critical components included: FO purifier, FO supply pump, generating engines 1-3, main engine thrust bearing, main engine turbocharger and LO purifier.

5.2 RAW DATA CONVERSION

From the critical equipment selection, it was seen that the main engine thrust bearing is one of the items under consideration. By applying the process described in chapter 3.3, the measurable variables for the thrust bearing were identified. These parameters were:

- ME Thrust Bearing LO Outlet Temperature [°C]
- ME Rotational Speed [rpm]

From these parameters, the LO temperature is a direct indicator of the bearing's condition. On the other hand, the ME rotational speed influences the bearing in an indirect way.

For the FES, the types of the membership functions selected were the best fit for the specific problem. The membership functions for the temperature and risk and show in figure 5 and figure 6 respectively. Should be specified that optimum temperature for the thrust bearing was 62°C, according to OEMs. Also, upper-temperature limits were obtained from the OEMs.

The FES correlated each temperature reading with the OEM limits. After that, a risk index value was produced. The resulting index was then adjusted based on the engine's rotational speed (rpm). For that purpose, a nominal operating point of 90rpm was adopted based on the ship's operating profile. For example, a high LO temperature at the nominal rpm would indicate an increase of the risk index. On the other hand, an increased temperature at high rpm would partially justify the high temperature (figure 7).

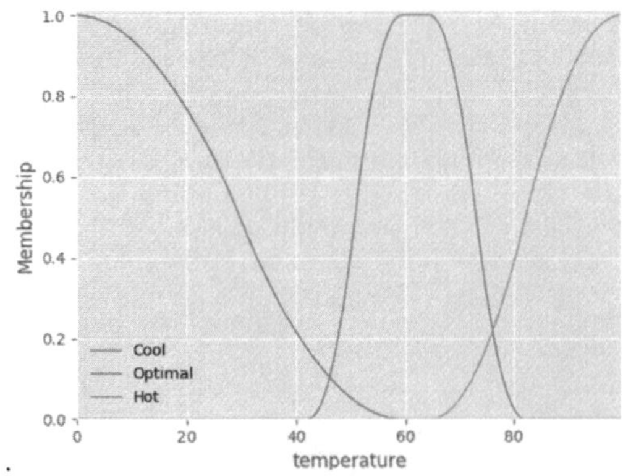

Figure 5: Temperature membership function

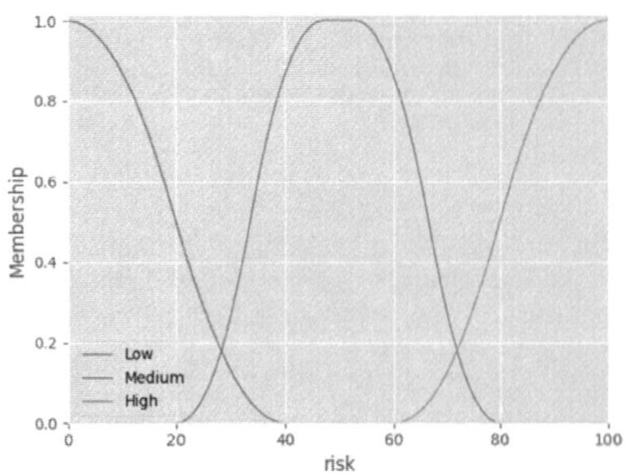

Figure 6: Risk Membership function

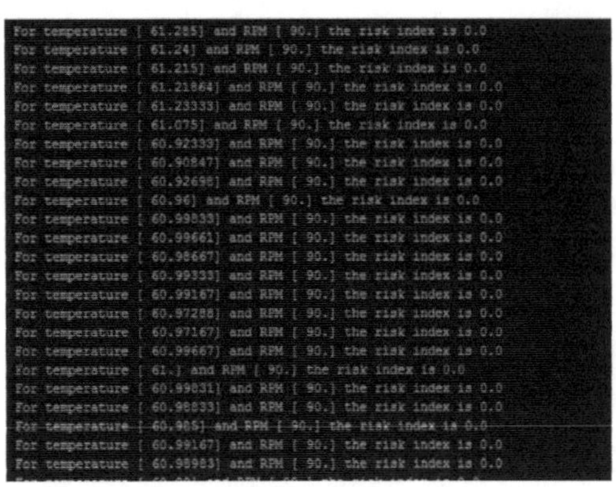

Figure 7: Sample Risk Index Result

6. CONCLUSIONS

Linking maintenance with the actual condition of the equipment is very important. For that purpose a clear strategy must be developed that dictates which items to monitor and which parameters to measure. Advancements in the fields of data mining and AI must be adopted with caution, while following international standards.

This paper uses established reliability assessment techniques (DFTA) in conjunction with modern data mining approaches (k-means clustering) to develop a structured tool for the identification of critical components. Following that, a methodology that assesses the condition of the equipment is developed. This methodology is based on modern AI tools and engineering expertise from OEMs.

Future work of this paper would include the examination dynamic aspects of the system with the end goal of making more accurate reliability predictions.

7. ACKNOWLEDGEMENTS

The work presented in this paper is partially funded by ISEMMS project. ISEMMS Project has received research funding from the United Kingdom's Innovate UK Programme. This publication reflects only the authors' views and the United Kingdom is not liable for any use that may be made of the information contained within

8. REFERENCES

1. BENGTSSON, M., KURDVE, M., 'Machining Equipment Life Cycle Costing Model with Dynamic Maintenance Cost', *23rd CIRP Conference on Life Cycle Engineering*, 2016.

2. STOPFORD, M., 'Maritime Economics 3e', *Routledge*, 2009.

3. WANG, Y., DENG, C., WU, J., WANG, Y., XIONG, Y., 'A corrective maintenance scheme for engineering equipment', *Engineering Failure Analysis,* 2014

4. FUMAGALLI, L., MACCHI, M., GIACOMIN, A., 'Orchestration of preventive maintenance interventions', *IFAC PapersOnLine*, 2017

5. LAZAKIS, I., RAPTODIMOS, Y., VARELAS, T., 'Predicting ship machinery system condition through analytical reliability tools and artificial neural networks', *Ocean Engineering*, 2017

6. TOURET, T., CHANENET, C., VILLE, F., LALMI, M., BECQUERELLE, S., 'On the use of temperature for online condition monitoring of geared systems – A review', *Mechanical Systems and Signal Processing*, 2018

7. ACCORSI, R., MANZINIA, R., PASCARELLA, P., PATELLAB, M., SASSIA, S., 'Data mining and machine learning for

condition-based maintenance', *27th International Conference on Flexible Automation and Intelligent Manufacturing*, 2017

8. EMSA, 'Annual Overview of Marine Casualties and Incidents 2016', *EMSA*, 2016

9. MCA, 'The ISM Code - Instructions for the Guidance of Surveyors', *MCA,* 2014

10. NGUYEN, K., DO, P., GRALL, A., 'Multi-level predictive maintenance for multi-component systems.', *Reliability Engineering and System Safety,* 2014

11. MOBLEY, K., HIGGINS, R., WIKOFF, J., 'Maintenance Engineering Handbook 7e', *McGraw Hill*, 2008

12 DIKS, K., 'Establishment of a novel predictive reliability assessment strategy for ship machinery', *University of Strathclyde*, 2017

13 KABIR, S., 'An overview of fault tree analysis and its application in model based dependability analysis', *Expert Systems with Applications,* 2017

14 RAO, D., GOPIKA, V., RAO, S., KUSHWAHA, H., VERMA, A., SRIVIDYA A., 'Dynamic fault tree analysis using Monte Carlo simulation in probabilistic safety assessment', *Reliability Engineering and System Safety,* 2009

15 ESPIRITU, J., COIT, D., PRAKASH, U., 'Component criticality importance measures for the power industry', *Electric Power Systems Research,* 2007

16 STRACHAN, M., BRUCE, S., MCARTHUR, S., 'Practical applications of data mining in plant monitoring and diagnostics', *IEEE,* 2007

17 OLSON, D., DELEN, D., 'Advanced Data Mininig Tenchniques', *Springer,* 2008

18 GIUDICI, P., FIGINI, S., 'Applied Data Mining for Business and Industry 2e', *Wiley*, 2009

19 NERURKAR, P., SHRIKE, A., CHANDANE, M., BHIRUD S., 'Empirical Analysis of Data Clustering Algorithms', *Procedia Computer Science*, 2018

20 KUSHWAHA, N., PANT, M., KANT, S., KUMAR, V., 'Magnetic optimization algorithm for data clustering', *Pattern Recognition Letters,* 2017

21 PEETERS, J., BASTEN, R., TINGA, T., 'Improving failure analysis efficiency by combining FTA and FMEA in a recursive manner', *Reliability Engineering and System Safety, 2018*

22 HAIGH, J., 'Probability Models 2e', *Springer,* 2013

23 POLI, J., BOUDET, L., 'A fuzzy expert system architecture for data and event stream processing', *Fuzzy Sets and Systems,* 2017

24 CATTANEO, M., 'The likelihood interpretation as the foundation of fuzzy set theory', *International Journal of Approximate Reasoning*, 2017

25 TIMOTHY, R., 'Fuzzy Logic with Engineering Applications 2e', *Wiley,* 2004

9. AUTHORS BIOGRAPHY

Michail Cheliotis holds the current position of Research Assistant and Ph.D. student at the Department of Naval Architecture, Ocean and Marine Engineering at the University of Strathclyde. He is responsible for conducting research and contributing to EU and UK projects relating to ship reliability, maintenance and safety.

Iraklis Lazakis is a Senior Lecturer at the department of NAOME of University of Strathclyde. His research interests include ship systems condition monitoring, systems maintenance and reliability, ship operations, risk analysis tools and methodologies, system criticality assessment, shipyard manufacturing and productivity, and asset management.

AUTONOMOUS MARINE NAVIGATION IN GLOBAL NAVIGATION SATELLITE SYSTEM (GNSS) DENIED ENVIRONMENTS

Mark R. Carter and **Geraint R. West,** Sonardyne International Limited, UK

SUMMARY

Recent Global Navigation Satellite System (GNSS) jamming and spoofing events emphasise the current dependence of shipping on GNSS and highlight the challenges of establishing an autonomous ship's position reliably in extreme and complex marine environments such as shipping channels, harbours and offshore energy installations. In response to these challenges, an Innovate UK funded project, "AutoMINDER" is seeking to integrate available positioning technologies with new sensor types and new integration methods, while establishment of a standardised integration architecture that supports easy adoption is proposed. This added redundancy, diversity and support for graceful degradation to fail-safe navigation is applicable to a number of unmanned and autonomous ship operations.

1. INTRODUCTION

Fault tolerant, high availability navigation systems used in the oil and gas industry today can already be considered "smart" and are an enabling technology for increased levels of vessel autonomy, which share requirements for high integrity navigation. However, with up to 200 people on-board, a typical offshore drilling rig is evidently far from being unmanned and still relies on human intervention when automatic fault detection and decision making are defeated.

The development of future system capability is moving towards both higher levels of automation and the enabling of remote access for troubleshooting, support and decision making activities. Fortunately many of the building blocks that achieve these developments already exist on the latest generation of offshore oil and gas vessels.

2. IMPACTS OF POSITIONING AND TIMING INTERFERENCE

Most modern infrastructure is dependent on GNSS for high precision positioning and timing. While precise GNSS timing has become fundamental to a diverse range of applications ranging from stock exchange transactions to cell phone tower synchronisation, GNSS functionality is now core to most aspects of marine traffic management. Indeed GPS jamming trials off Flamborough Head by the General Lighthouse Authority in collaboration with the Defence Science and Technology Laboratory [1] have demonstrated significant secondary impacts such as corrupted Automatic Identification System data, in addition to the direct denial of navigation onboard a single vessel: In short, denial of GNSS even for short periods, has the potential to create chaos in busy shipping areas.

GNSS signals are inherently low in power due to the distances the signal travels from space vehicles. This can lead to these being blocked or attenuated in situations where there the view of the sky is masked (typically below 10° elevation). In the marine environment, this can typically occur when operating close to offshore platforms or high buildings, near mobile cranes, surveying in inshore areas and around offshore wind farms. The consequence is often a temporary loss or distortion of GNSS positioning for a few seconds.

A simple noise jammer can transmit spurious signals at or near the GPS frequencies (which are well known); these signals can be pure white noise or modulated with a random code and can be broad or narrow band in nature. As well as simple noise jamming there is 'smart jamming', which uses knowledge of the GNSS signal structure to cause disturbance to a receiver. Three jamming methods use this form; these are known as code jamming, spoofing and meaconing. In one of the most publicised instances of this during June 2017, at least 20 ships in the Black Sea reported an apparent GPS spoofing attack, which placed these ships ca. 25 NM away from their actual positions inland (Figure 1) [2].

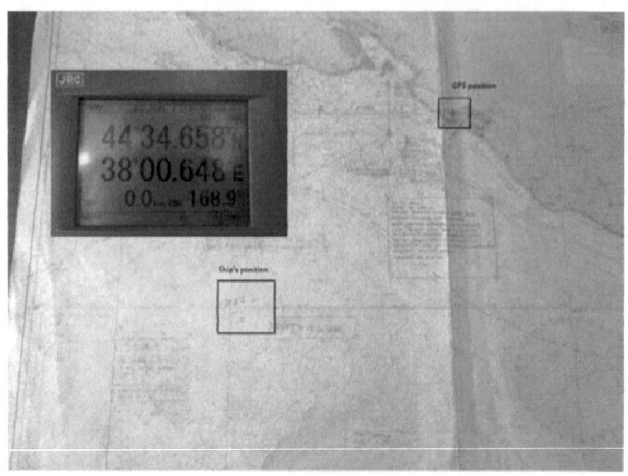

Figure 1: Apparent GPS spoofing off Novorossiysk

GNSS signals are also vulnerable to an ionospheric disturbance (caused by anomalous solar activity) known as scintillation, which causes amplitude and phase variations on signals from GNSS satellites when they cross the ionised upper atmosphere. Effects range from degradation of positioning accuracy to the complete loss of signal tracking.

3. CURRENT STATE OF THE ART DYNAMIC POSITIONING REFERENCE SYSTEMS

3.1 OVERVIEW

IMO 645 rules require at least three independent position references to be available on Class 2 or 3 Dynamically Positioned (DP) vessels. Typically, two or more Differential GNSS (DGNSS) systems are used alongside one or more Long and Ultra Short Baseline Acoustic (L/USBL) reference systems and/or local reference systems such as radar or laser relative position sensor. In order to claim independence, the DGNSS systems are usually manufactured by different companies and with differential corrections delivered by various means. A conventional DP system is shown in Figure 2.

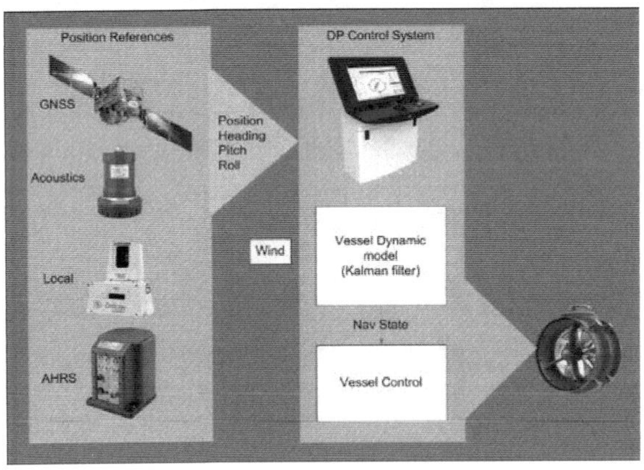

Figure 2: Conventional DP system

3.2 VULNERABILITIES

While using two or more DGNSS systems with acoustic and local references, meets requirements for independence, a conventional DP system is still vulnerable to a number of impacts:

- GNSS positioning vulnerable to systematic errors and signal outages as described in Section 2 above.
- Acoustic based L/USBL systems are highly repeatable and robust, but can be interfered with by environmental factors such as under-hull aeration, episodic increases in noise interference sources and physical masking
- Local reference systems are reliant on identifying the targets they need in their field of view: These are often either not available or cannot be easily installed offshore.

As a result, no DP manufacturer can provide position data with 100% availability and the consequences of the combined effects and constraints described above have been attributed to a number of station keeping incidents over recent years in the oil and gas industry. The lessons learned in this industry have highlighted the need to improve the robustness of position measurement equipment. These developments are of increasing importance as levels of autonomy increase and are also potentially applicable to the maritime industry in general.

4. ROBUST NAVIGATION IN COMPLEX ENVIRONMENTS

4.1 TECHNICAL APPROACH

The challenge of establishing position reliably in extreme and challenging marine environments such as shipping channels, harbours or at offshore energy installations is being addressed as part of the Innovate UK AutoMINDER (Autonomous MarIne Navigation in Denied EnviRonments) project. AutoMINDER is extending the available positioning technology by incorporating new sensor types and new integration methods as shown in Figure 3.

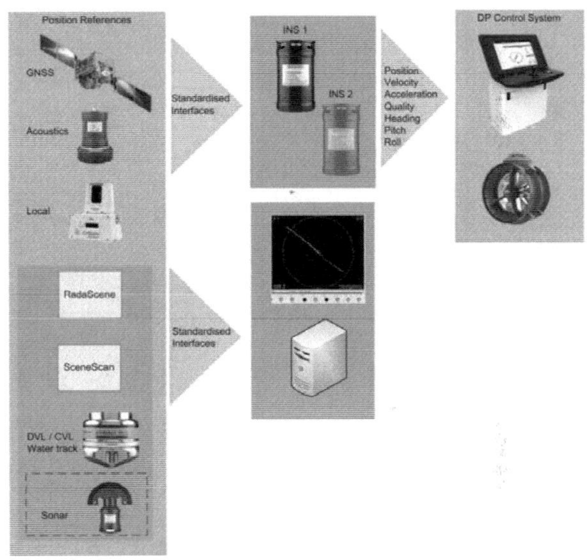

Figure 3: AutoMINDER project

The innovation in AutoMINDER is highlighted above in Figure 3. The project builds on the current state of the art in positioning sensor systems (GNSS, Inertial Navigation Systems (INS) and Doppler Velocity Logs (DVL)) by integrating Correlation Velocity Logs (CVL), water velocity tracking and multi-sensor range and bearing measurements from target and target-less optical and radar based surface sensors. This will add integrity due to redundancy and diversity of sensor type and extend operation to include GNSS denied environments.

4.2 INERTIAL NAVIGATION

Sonardyne International is focusing on an inertial navigation sensor with integrated water-track DVL and CVL to constrain drift.

Inertial navigation is the computation of velocity, position and orientation from measurements of acceleration and rotation. Acceleration (velocity change) is measured by an orthogonal triad of accelerometers. Rotation (angular change) is measured by an equally orthogonal triad of gyroscopes. The composite mechanical structure comprises the Inertial Sensor Assembly (ISA). An Inertial Measurement Unit (IMU) is composed of the ISA and a processor that performs various forms of compensation, including factory calibration coefficients. The outputs of an IMU are compensated delta-angles and delta-velocities, typically at an update rate of 100-200Hz.

An INS is based on the dead-reckoning principle: INS velocity is computed from an initial value by integration of acceleration over time. Similarly a position vector is computed from velocity and an orientation vector (attitude and heading) is computed from integration of measured rotation (angular change). It is self-evident that a fixed error (bias) in acceleration when integrated will cause an error in velocity that grows proportionally with time. In turn, the velocity error when integrated causes a quadratic growth in the positional error, sometimes known as drift. The drift of a typical commercial aircraft INS is of the order of one Nautical Mile per Hour. This is adequate for guidance from one airport to another (e.g. crossing the Atlantic) but hardly useful for navigation of slow moving marine vehicles over periods of weeks and months or where high accuracy positioning is required.

Improved performance can be achieved by Aiding the INS (AINS) with measurements from one or more external sensors as shown in Figure 3; The AINS algorithm is a Kalman filter that determines INS errors from external sensor measurements. The difference between a measurement (observation) and its computed value (based on the INS) is fed into the Kalman filter and is used to refine its state vector and covariance matrix. The state vector is composed of INS errors in position, orientation and velocity, and often also key inertial sensor errors such as gyroscope and accelerometer bias. Errors in external sensors can also be modelled for improved performance.

Due to the need for external aiding, an INS is not a completely independent reference; however, the combination of GNSS positioning and INS does provide a complementary solution that increases the integrity of a vessel's navigation system, which has consequently reduced the number of station keeping incidents. The following important advantages are also realised:

- "Ride Through Capability"- resilience to short term outages in the acoustic reference or GNSS due to noise increase and aeration.

- Faster update rate (1Hz – 5Hz) and high precision improves DP weighting that increases the probability of detecting GNSS disruptions.

4.3. DOPPLER AIDED INERTIAL NAVIGATION

Measurement of speed over ground using a Doppler velocity log offers an additional navigation input to increase reliability. When combined with INS it is possible to accurately position a vessel for many minutes in the absence of GNSS.

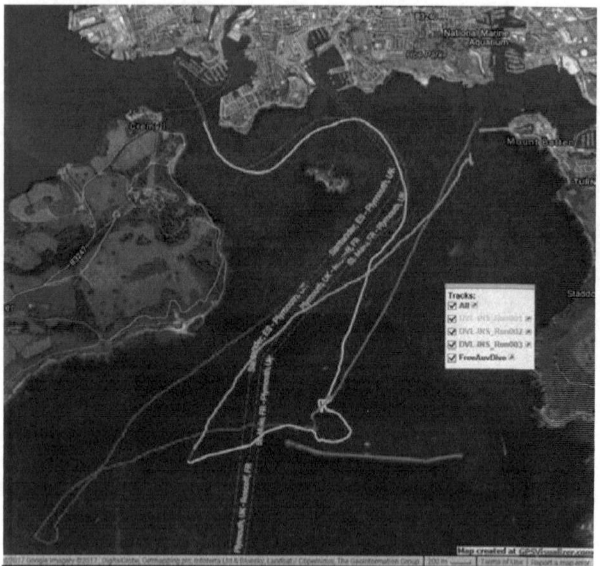

Run #	ΔTime [min]	Distance travelled [km]	Radial Error [m]	[% Error] Distance
1	74.00	6.42	1.77	0.028
2	33.50	3.72	1.24	0.033
3	49.70	3.05	1.43	0.047
Mean	**52.40**	**4.40**	**1.48**	**0.036**

Figure 4: DVL aided INS position error

Figure 4 above shows a mean 52.4 minute, 4.4km shallow water manoeuvre in the absence of GNSS. The vessel is positioned using just DVL and inertial navigation with a maximum absolute position error of 1.48m (0.036%) over the time and distance.

4.4 LASER AND RADAR AIDED INERTIAL NAVIGATION

Guidance Marine is focusing on the development of environmentally referenced positioning sensors to provide additional integration and redundancy to aid the INS and constrain drift.

All current commercially available local positioning surface reference sensors cannot function without dedicated prepositioned targets. These targets need to be installed on a structure to enable the vessel to approach or control is relative position safely (see Figure 5). Despite the continuous development of target hardware and software, this limitation critically limits relative position measurements to only operational areas with available targets, which represents a huge problem to the marine industry in terms of safety and ownership.

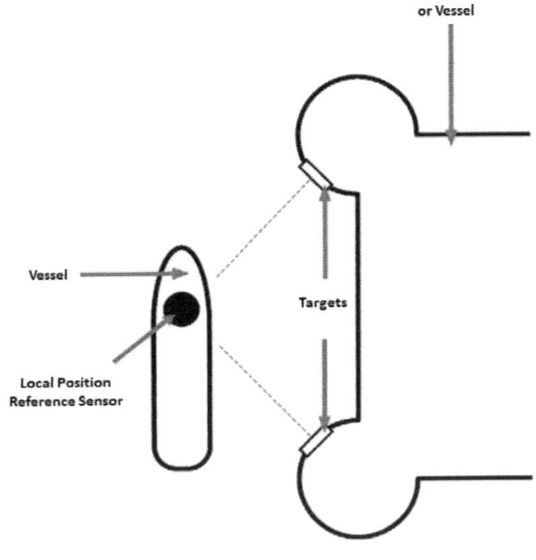

Figure 5: Set up of local position reference targets

The locations of targets are limited by available architecture and consequently not always optimally sited. This can be compounded if targets are moved by personnel, are damaged, become detached, require maintenance or (in some cases) need re-charging. There are also significant set up and maintenance costs associated with these targets. The requirement for locating the targets adds commercial complexities, in terms of installing, maintaining and liability. Locating targets on unmanned installations is a particular challenge.

Consequently the potential to realise autonomous operations and improve the reliability of navigation in GNSS denied environments requires the availability of environmentally referenced sensors. The innovative technology proposed by Guidance uses laser or radar backscatter from infrastructure as a target and the latest state of the art developments in intelligent Simultaneous Localisation and Mapping (SLAM) algorithms to determine how the scene changes. This system, SceneScan, provides a view of the environment around the vessel (Figures 6a and 6b). [3]

This sensing modality enables simultaneous mapping of an area and positioning of the vessel relative to the generated map. The extracted range and bearing information will be used to aid the INS and constrain drift. In combination with the development of proprietary laser optic, microwave transmission and mechanical systems, it allows position referencing to be achieved in a similar way to that used in autonomous terrestrial robots and vehicles, but in the harsh marine operating environment.

Figure 6a: SceneScan on sea trial - Gulf of Mexico

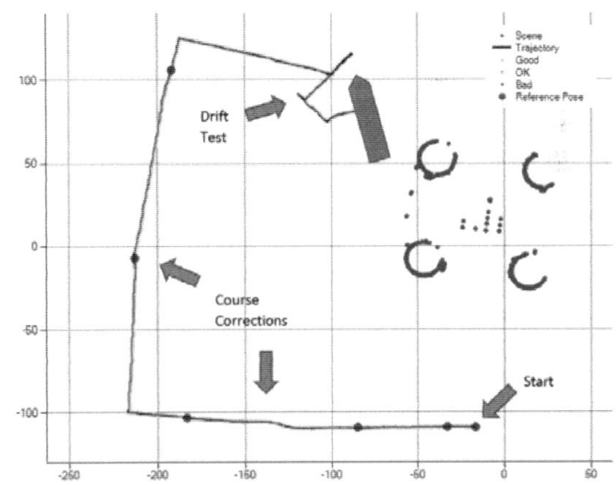

Figure 6b: 2D map tracked by SceneScan

5. NAVIGATION SENSOR INTEGRATION

Navigation sensor type and specification are likely to vary according to vessel size and mission. A Navigation Sensor Suite, including existing and new sensor types with improved sensor integration will be developed to deliver the required position accuracy and availability appropriate to the autonomous vessel type, operational mission and voyage segment (i.e. docking, harbour, shipping channel, oceanic, or station keeping). This is particularly required for autonomous operation in the absence of direct human control.

As shown in Figure 8 below, design modularity including plug and play sensors with standardised machine to machine interfaces will add operational efficiency. Common data fusion, remote command and control, diagnostics, data management and displays will simplify oversight by human operators in situ or ultimately remote from the vessel.

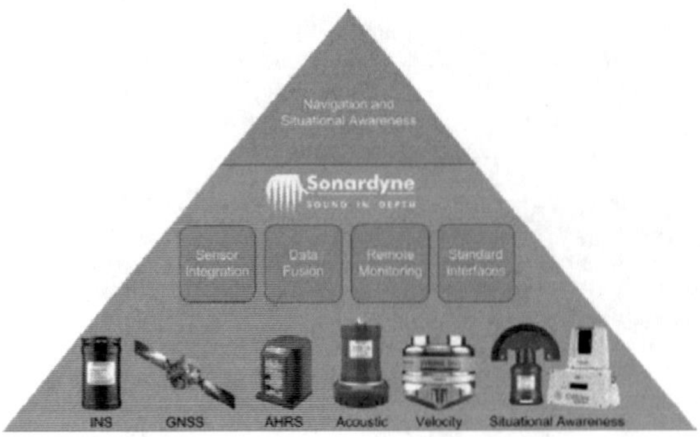

Figure 8: Scalable sensor integration

6. COMMAND AND CONTROL

Given that much of the data required to support high integrity navigation and increased levels of autonomy are already available on-board from existing sensors a further consideration is the value of remotely monitoring that data for trouble shooting, support and decision making operations. Various stakeholders have different requirements:

6.1 SUPPORTING DATA

Extensive quality metrics related to DP position such as measurement noise and associated error metrics are available which, if recorded and processed through a data analytics approach, allow sensors to become truly smart and self-learning. These sensors store both a priori and posteriori performance, which has the potential to provide a number of operational advantages:

6.1 (a) Optimised System Performance

Since sensor data can be accessed not just on an instantaneous basis, but also averaged for specific operational periods with particular environmental conditions or geographic locations. This quality history potentially enables the DP system to be smarter about sensor weighting and selection.

System performance and position integrity can also be remotely monitored over time; this would enable degraded or suboptimal performance due to inappropriate system configuration or environmental factors to be identified ashore in near-real time. The vessel's system can consequently be optimised for performance in particular scenarios or following environmental and / or operational changes.

6.1 (b) Remote Condition Based Monitoring

A wide range of sensors can be remotely monitored by support centres worldwide over time to detect deterioration in performance and potential need for maintenance, including:

- The long term trend of GNSS signal-to-noise (SNR) ratio on a vessel can be monitored to identify any problems with antenna condition or installation.

- USBL calibration status and transceiver performance can be measured and any degradation highlighted in advance of any material loss in performance. Transducer inspection and cleaning can consequently be carried out during convenient periods of downtime rather than as unscheduled maintenance.

- Transponder battery status can be monitored remotely and replacements ordered by a support centre to coincide with port calls and supply vessels.

- The performance of solid state sensors in ships' gyros, motion reference units and IMU can be monitored to detect changes in bias and drift rates indicating the need for re-calibration.

A move away from time based maintenance would offer significant logistics savings and provide fleet wide sensor performance statistics.

6.2 REMOTE CONTROL ARCHITECTURES

It would be inefficient for individual sensor and sub-system manufacturers to implement their own end-to-end remote monitoring solution including on-board data collection, communication system and shore base data centre functions. Fortunately, the major DP manufacturers and some vessel owners existing infrastructure has open interfaces that can be used by Third parties. However, the details of interfaces, networks and data centres are different for each provider even where the overall aims and approach are similar.

At the moment a rigid architecture may not work, and some flexibility is needed in the absence of relevant industry standards. In order for the full value to be realised, standards need to be developed that will allow sensors, subsystems, vessels and data centres to share data effectively and securely.

7. CONCLUSIONS

Autonomous vessels navigating in complex marine environments such as shipping channels, harbours or close to offshore energy installations need to behave in a safe and predictable manner when external GNSS-based positioning services are denied.

A Navigation Sensor Suite (NSS) including existing and new sensor types with improved sensor integration is being developed as under the AutoMINDER project to deliver the required position accuracy and availability appropriate to the autonomous vessel type, industrial mission and voyage segment (e.g docking, harbour, shipping channel, oceanic, or station keeping). Design modularity including plug and play sensors with standardised machine to machine interfaces will add operational efficiency. Common remote command and control, diagnostics, data management and displays will simplify oversight by human operators in situ or ultimately remote from the vessel.

8. ACKNOWLEDGEMENTS

The help from Martin Schaefer, Guidance Marine, UK, in preparing this paper is gratefully acknowledged.

9. REFERENCES

1. GRANT, D., WILLIAMS, P., SHAW, G., DE VOY, M. & WARD, N., 'Understanding GNSS availability and how it impacts maritime safety', International Technical Meeting of the Institute of Navigation, January 24 – 26, 2011.

2. https://www.rin.org.uk/newsitem/4969/GPS-Spoofing-in-Black-Sea, 12 July 2017.

3. SANDERSON, D., 'Reducing the Interdependence Failure Opportunities – Environmental Referencing ("Targetless") Sensors for DP Applications', Guidance Marine, March 2016.

10. AUTHORS BIOGRAPHIES

Mark Carter was the Business Manager for Vessel Equipment at Sonardyne International Limited between 2011 - 2017. He has 20 years of navigation systems experience across the maritime and aviation sectors. Previously, Mark worked at QinetiQ as engineering manager for the navigation group. This involved working for military and commercial customers to develop GPS anti-jam systems and high integrity aided inertial navigation systems. Mark has a background in Systems Engineering and a BSc in Surveying from the University of Newcastle-Upon-Tyne.

Geraint West is Business Manager for Ocean Science at Sonardyne International Limited. Previously, he was Director National Marine Facilities at the National Oceanography Centre, during which time he set up its new Marine Autonomous and Robotic Systems facility and commissioned two new ocean research ships, RRS James Cook and RRS Discovery. Originally trained as a hydrographic surveyor, Geraint served in the Royal Navy for 15 years, as well as 3 years with John E. Chance (Fugro), managing the US Army Corps of Engineers' airborne laser hydrographic survey system.

IMPLEMENTING UNSUPERVISED LEARNING ALGORITHM FOR MARINE ENGINE DATA CLUSTERING APPLICATIONS

Y Raptodimos and **I Lazakis**, University of Strathclyde, UK

SUMMARY

Data preparation and processing is of great importance in a ship condition monitoring tool, as inaccurate and misinterpretation of data can significantly affect the condition monitoring accuracy and performance. Data for performance parameters related to the case study of a Panamax container ship main engine are clustered using an artificial neural network, the Self-Organizing Map (SOM). Neighbouring clusters are compared through a distance metric to examine the existence of data similarities. Additionally, the SOM has a supplementary functionality of identifying data clusters exceeding thresholds, consequently providing diagnostics connected to a Failure Mode and Effects Analysis (FMEA) for the main engine, providing valuable insight and information regarding potential faults. The SOM model is validated through actual data extracted from the case study. Moreover, simulated data representing data exceeding alarm levels for the engine fuel oil system demonstrate the capabilities of the SOM clustering process in combination with the associated FMEA results.

1. INTRODUCTION

Industries across the spectrum are coming to terms with the fundamental changes that technological advances in areas such as artificial intelligence, robotics and automation will have. The marine industry is facing a tipping point, where technological developments such as autonomous and smart ships are turning into reality. Concurrently, the sector is facing unprecedented cost pressures and a stricter regulatory backdrop, as environmental restrictions persist. Smart shipping is the inclusive term coined by the industry to describe the digital technologies available for determining and optimising operational safety and efficiency [1]. Tightening margins and the affordability and availability of computing power are combined to form opportune conditions for the adoption of smart shipping. As such, data analytics and condition monitoring tools associated with predictive maintenance tasks, are popular options for enhanced commercial efficiency and safety.

Technological advances and high cost of ownership have resulted in considerable interest in advanced maintenance techniques [2]. Though the industry is still predominantly reliant on a time-based, prescriptive approach to maintenance, there are several factors challenging the long-held norm. The increasing complexity of shipboard systems, heightened expectation and competitive needs as to ship and plant availability and efficiency and the influence of the data revolution on vessel operations, favour a properly structured Condition Based Maintenance (CBM) regime [3]. British Standard [4], define CBM as the maintenance policy carried out in response to a significant deterioration in a machine as indicated by a change in a monitored parameter of the machine condition. The heart of CBM is condition monitoring. Condition monitoring is the process of monitoring parameters expressing machinery condition, interpreting them for the identification of change which could be indicative of developing faults. Jardine et al. [5] and Kobbacy and Murthy [6] divide CBM into three main steps: data acquisition, data processing and maintenance decision making. Data pre-processing and post-processing is of great importance in a ship condition monitoring software tool, as misinterpretation of data can significantly affect the accuracy and performance of the predictions made.

In this respect, the question of how much data, which data, and how often this should be collected and how has also arisen, as although companies adopt CBM schemes, there seems to be an issue in processing, analysing and utilising the recorded operational information. Nowadays, data collected by sensors used on ship machinery incorporate an enormous amount of measurement instrumentation, including temperature, pressure, flow, vibration and current sensors. To explore, extract, and generalize inherent patterns in spatiotemporal data sets, clustering algorithms are indispensable [7]. Consequently, this paper focuses on presenting the employment of Self-Organizing Maps in order to cluster multidimensional data. A Self-Organizing Map (SOM) is a type of Artificial Neural Network (ANN), trained through unsupervised learning for transforming an incoming signal patter of arbitrary dimension into a one or two-dimensional discrete map, and to perform this transformation adaptively in a topologically ordered fashion.

The SOM is applied in the case study of a two-stroke marine diesel engine as part of a data processing and preparation stage of a condition monitoring tool. Furthermore, different clusters that contain analogous data are grouped by similarity using a distance metric. The case study presented in this paper demonstrates the application of the SOM for clustering monitored data of key performance parameters for a main engine cylinder based on Original Equipment Manufacturer (OEM) thresholds. Moreover, the SOM has an additional functionality by identifying clusters of data exceeding OEM thresholds, thus acting as a diagnostic tool which is connected to a Failure Mode and Effects Analysis

(FMEA) for the main engine in order to provide useful insight and information regarding possible failure events and causes of failed main engine systems/components.

This paper is organized as follows. First, Section 2 summarises the existing literature related to this work, followed by the demonstration and description of the methodology developed in Section 3. Section 4 presents the case study for the ship main engine through which the methodology is applied alongside the obtained results. Finally, Section 5 establishes the conclusions of this paper providing remarks on the work achieved and directions for future research and development.

2. BACKGROUND

2.1 SELF-ORGANIZING MAP (SOM)

The SOM is a neural network concept developed by Kohonen [8]. It forms a one or two-dimensional presentation from multi-dimensional data. The topology of the data is kept in the presentation such that data vectors, which closely resemble one another, are located next to each other on the map. A neural network can be defined according to Principe et al. [9] as distributed, adaptive, generally nonlinear learning machines built from many different processing elements that receive connections from other processing elements and/or itself. In contrast to traditional methods, such as principal component analysis, the SOM can also be created from highly deviating, nonlinear data. Therefore, the SOM is a flexible, unsupervised neural network for data analysis and clustering [7].

Unsupervised learning is a type of machine learning algorithm used to draw inferences from datasets consisting of input data without labelled responses. The most common unsupervised learning method is cluster analysis, which is used for exploratory data analysis to find hidden patterns or grouping in data. In unsupervised learning, there are no predetermined classes, thus no labelled data are available, and the goal of clustering is to separate a finite, unlabelled data set into a finite and discrete set of hidden data structures. A direct reason for unsupervised clustering is the need to explore the unknown nature of the data that are integrated with little or no prior information [10]. Figure 1 illustrates the basic structure of the SOM. It is a class of neural networks with neurons arranged in a one or two dimensional structure and trained by an iterative unsupervised or self-organising procedure [11]. They find clusters in the data by evaluating neighbourhood measures and employing competitive learning strategies. The output neurons of the network compete among themselves to be activated or fired, with the results that only one output neuron, or one neuron per group is on at any one time. An output neuron that wins the competition is called a winner-takes-all neuron or winning neuron. Every data item is mapped into one point (node) in the map and the distances of the items in the map reflect similarities between the items [12].

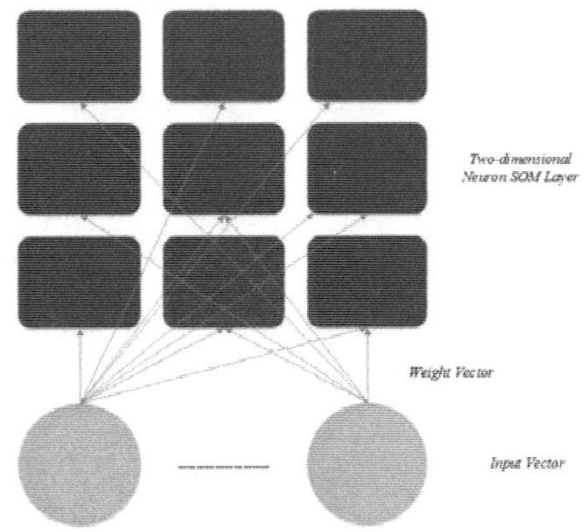

Figure 1: Self-Organizing Map Structure

During training, the SOM forms an elastic net that folds towards the space formed by the input data. Data points lying near each other in the input space are mapped onto nearby map units. After the input data are normalized, the SOM is trained iteratively. In each training step, one sample vector X from the input data set is chosen randomly, and the distance between it and all the weight vectors of the SOM, which is originally initialized randomly, is calculated using some distance measure, such as Euclidian distance. The Best Matching Unit (BMU) is the map unit whose weight vector is closest to X. After the BMU is identified, the weight vectors of the BMU as well as its topological neighbours are updated so that they are moved closer to the input vector in the input space.

The SOM can be interpreted by labelling the units according to input vectors, whose type or operation state is known. Unlike networks based on supervised learning which require that target values corresponding to input vectors are known, the SOM can be used for clustering data without knowing the class membership of the input data. It can be used to detect feature inherent to the problem. Therefore, the SOM is an effective tool for condition monitoring and system degradation detection.

2.2 FAILURE MODES AND EFFECTS ANALYSIS (FMEA)

FMEA provides a systematic method for organizing the study of a particular system or process in terms of failure analysis. The aim of FMEA is to review the system in order to provide details on how to identify failures and their causes as well as determine the end results of the failures occurring. Thus, FMEA is a formalized method to consider all components, their functions, failure modes and system failures [13]. It involves reviewing as many components, assemblies and subsystems as possible to identify possible failures modes and causes and effects of such failures. FMEA can be applied in a bottom-up approach which assists in mapping the overall failure potential of the system. This technique is most suited for the risk assessment of mechanical and electrical systems and the approach can be either quantitative or qualitative. According to Ben-Daya and Knezevic [14] FMEA performs three functions. These are initially the identification and recognition of potential failures including their causes and effects, the evaluation and prioritization of identified failure modes and the identification and suggestion of actions to eliminate or reduce the chance of the potential failures occurring.

The American Bureau of Shipping has released guidance notes [15] related to FMEA requiring the development and submission of FMEAs as part of Classification requirements for certain systems such as Dynamic Positioning systems, drilling systems, dual fuel diesel engines etc. The International Association of Classification Societies (IACS) [16] published recommendations for the FMEA process for diesel engines and reporting the FMEA process. To contribute and improve the ongoing efforts of Classification Societies and operators, Cicek and Celik [17] examined the application of FMEA in order to prevent and reduce the occurrence of crankcase explosion failure in order to improve machinery system reliability and enhance operational safety on board ships.

3. METHODOLOGY

3.1 METHODOLOGY OVERVIEW

The overview of the methodology suggested in this paper is demonstrated in Figure 2. At first, data collected from onboard ship sensors are used as input in the SOM. The SOM is initially trained, during which the map is defined and shaped based on the input data while the second phase automatically classifies new inputs into the clusters defined in the training stage. Subsequently, the SOM alongside the distance metric analyse the data from the ship sensors into cluster groups.

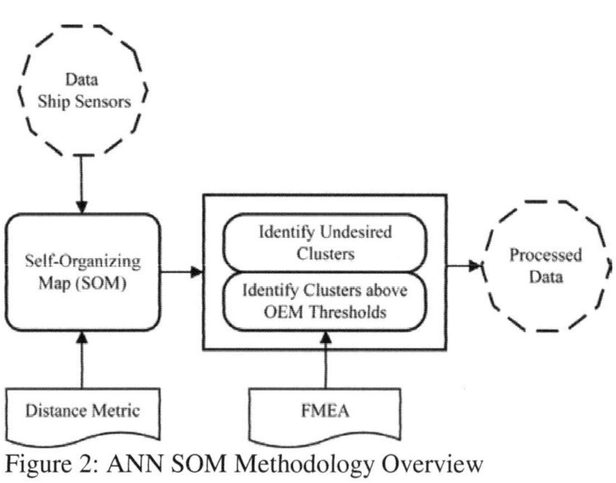

Figure 2: ANN SOM Methodology Overview

The clusters can represent healthy data, meaning data that are within the OEM thresholds for the main engine, data exceeding thresholds and excessive data. In the case of data exceeding OEM thresholds, the identified clusters for the associated engine performance parameters are linked through a developed in-house MATLAB code with associated information contained in the main engine FMEA. In the case of excessive data, these are described as clusters containing data that do not reflect under any circumstance values representing normal engine operating conditions or values beyond alarm levels. It is assumed that these clusters are due to error in the data acquisition system or human error during the data recording and collection process. Finally, these undesired clusters can be removed, leading to the data being processed for further analysis in a condition monitoring software tool.

3.2 SOM & DISTANCE METRIC

3.2 (a) SOM Methodology

The SOM consists of an input and output layer. Inputs in the SOM are the input vectors containing data measurements of main engine performance parameters. Additionally, the topology of the SOM defines the number of neurons (clusters) and a distance function in order to obtain distances between the neurons given their positions. The output of the algorithm is the number of neurons the input data has been assigned to. The neurons in the layer of a SOM are arranged in physical positions originally, according to a topology function, arranging the neurons either in a grid, hexagonal or random topology.

The network is trained using the batch unsupervised weight/bias training algorithm. Batch training of a network proceeds by making weight and bias changes based on an entire set (batch) of input vectors. Incremental training changes are applied to the weights and biases of the network after presentation of each individual input vector. Initially, the input data is clustered in the SOM in order to produce the clusters. Afterwards, the clusters obtained from the SOM analysis can be further distinguished into groups by applying the Euclidean distance metric amongst the centres of the SOM clusters. Clusters can be categorized under the same group based on their closeness.

3.2 (b) Distance Metric

Cluster centres with small Euclidean distances between them contain similar data and can be contained under one cluster group. A Euclidean space allows the representation of a cluster by its centroid or average of the points in the cluster The Euclidean distance between two points p and q is the length of the line segment connecting them. In Cartesian coordinates, if p and q are two points in Euclidean n-space, then the distance d from p to q or vice versa is given by the Pythagorean formula [10]. Interclustering distances are defined by calculating the Euclidean distance between the SOM clusters and selecting the clusters with the shortest distance. Stopping can be achieved by considering the number of clusters that should be in the data or when the best combination of existing clusters produces a cluster that is inadequate, pre-defined by the user or when the Euclidean distances exceed a threshold [18].

In this paper, the Euclidean distances of the SOM cluster centres are imported into a custom algorithm using a combination of logical operators and conditional statements and expressions in order to compare cluster distances and search for the clusters that have the shortest distance between them. The algorithm searches for clusters containing similar data, starting from cluster centres that have the closest distance towards larger ones by an increment of 50%. The algorithm stops searching for close clusters if the distance is larger than 0.5, as in this case the cluster distances are large, and no similarities exist in the data of those clusters.

3.3 FMEA

Applying FMEA follows a series of successive steps such as the analysis of the examined system, list of systems and components, list of identified potential failures and their potential causes, detection techniques and the local and global effect the identified failures have to the examined system. The main aim of the FMEA is to review as many components, assemblies and subsystems as possible to identify possible failures modes and causes and effects of such failures. In the case in which the clusters from the SOM analysis represent main engine data exceeding OEM thresholds, the identified clusters are linked through a developed in-house code developed in MATLAB with associated information contained in the main engine FMEA for the relevant main engine system or component.

4. CASE STUDY

This section presents a description of the case study following the results obtained from training the SOM with the case study data. Moreover, a sample of the constructed main engine FMEA is displayed. Results are shown for two case study scenarios. The first scenario demonstrates actual results for input corresponding to healthy data, while the second scenario simulates data owing to a fault in the main engine fuel oil system. Finally, the trained SOM results are validated through data collected during an onboard measurement campaign.

4.1 CASE STUDY DESCRIPTION

Data from a Panamax container ship are used related to the two-stroke marine diesel engine. Table 1 displays the main particulars for the vessel case study.

Container Ship Main Particulars	
DWT (Summer)	50829 tons
Length Overall	260.00 m
Beam	32.00 m
Depth Moulded	19.30 m
Draft (Summer)	12.60 m

Engine Particulars	
Main Engine	HSD 8K90MC-C
Maximum Continuous Rating (MCR)	49680 BHP @ 104 RPM
Cylinders	8
Bore	900 mm
Stroke	2300 mm

Table 1: Ship main particulars

The performance parameters used for the case study are shown in Table 2.

System/ Component	Measurable Parameters
Cylinder	Scavenging Air Temperature (per cylinder) Exhaust Gas Outlet Temperature (per cylinder) Jacket Fresh Water Cooling Outlet Temperature (per cylinder)
Piston	Piston Cooling Lubrication Oil Inlet Pressure Piston Cooling Oil Outlet Temperature (per cylinder)
Fuel	Fuel Oil Inlet Temperature Fuel Oil Inlet Pressure
Air	Air Cooler Cooling Water Inlet Pressure
Main Lubrication Oil	Main Lubrication Oil Inlet Pressure Main Lubrication Oil Inlet Temperature
Thrust Bearing	Thrust Bearing Lubrication Oil Outlet Temperature

Table 2: Main engine measured parameters

4.2 SOM TRAINING

The SOM created for clustering the data consists of a 10-by-10 two-dimensional map of 100 neurons. The SOM is trained for a maximum of 1000 epochs. Prior to training, the collected monitored data are normalized in order to standardize the range of independent variables of data. Once training is complete, the multidimensional input data vectors have been assigned into clusters. The SOM topology after training is shown in Figure 3, in which the green dots represent the input data for the cylinder main engine parameters and the red dots represent the SOM neurons-clusters assigned to the data points while the blue lines connect each node of the map.

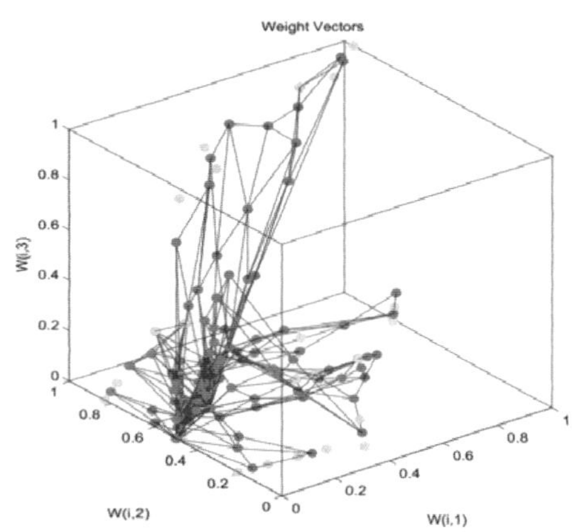

Figure 3: SOM topology after training

The clusters obtained after the training of the Self-Organizing Map alongside the data they contain are presented in Table 3.

Description	SOM Cluster
All monitored parameters in normal operating conditions	49, 77, 87, 99
Cylinder Scavenging Air Temperature above OEM threshold	51, 62, 92
Cylinder Scavenging Air Temperature excessive	61, 71, 81, 91
Cylinder Exhaust Gas Outlet Temperature above OEM threshold	80, 89
Cylinder Exhaust Gas Outlet Temperature excessive	79, 100
Cylinder Jacket Fresh Water Cooling Outlet Temperature above OEM threshold	26, 27, 86, 95
Cylinder Jacket Fresh Water Cooling Outlet Temperature excessive	5, 6, 16
Piston Cooling Lubrication Oil Inlet Pressure above OEM threshold	34, 35, 45
Piston Cooling Lubrication Oil Inlet Pressure excessive	28, 67, 96
Piston Cooling Oil Outlet Temperature above OEM threshold	59, 66, 75, 76
Piston Cooling Oil Outlet Temperature excessive	50, 55, 60, 65
Fuel Oil Inlet Temperature above OEM threshold	74, 82
Fuel Oil Inlet Temperature excessive	30, 40, 72, 73
Fuel Oil Inlet Pressure above OEM threshold	68, 69, 78
Fuel Oil Inlet Pressure excessive	10, 93, 94
Air Cooler Cooling Water Inlet Pressure above OEM threshold	29, 84, 97
Air Cooler Cooling Water Inlet Pressure excessive	20, 36, 37, 47
Main Lubrication Oil Inlet Pressure above OEM threshold	8, 9, 18
Main Lubrication Oil Inlet Pressure excessive	58, 85
Main Lubrication Oil Inlet Temperature above OEM threshold	19, 25, 52, 53
Main Lubrication Oil Inlet Temperature excessive	32, 33, 43
Thrust Bearing Lubrication Oil Outlet Temperature above OEM threshold	39, 88, 98
Thrust Bearing Lubrication Oil Outlet Temperature excessive	21, 31, 38, 41
All data excessive	1, 2, 3, 12, 13

Table 3: Self-Organizing Map Data Clusters

As an example, from Table 3, data corresponding to the fuel oil inlet temperature being above the OEM threshold is clustered into cluster 74 and 82. Moreover, not all 100 neurons of the SOM have been assigned to the input data, thus remaining inactive. This is due to the nature of the data used for the SOM training. However, the SOM algorithm treats all the neurons on the map as active and does learning accordingly. After the network training phase, the network is saved in order to carry out additional simulations using new data as input. To validate the network performance in clustering data successfully, new input data is used to simulate the ANN-SOM model.

4.3 MAIN ENGINE FMEA

Information collected for the development of the FMEA is based on Cicek and Celik [17], Turan et al. [19], Mokashi et al. [20], Emovon [21], INCASS EU project deliverables [22, 23] and experts opinion. Information was collected from three Class societies, two ship operators and two consultancy companies. Specifically, experts included IACS Classification Societies surveyors and researchers with many years of onsite experience. Moreover, experts from ship operators and marine consultancy companies included technical managers, superintended engineers, and Chief engineers sailing onboard the mentioned vessel. A sample of the constructed FMEA table is illustrated in Table 4.

Failed Item	Failure Event	Failure Cause	Local Effect	Global Effect
Cylinder Head	Cracked	Overheating	Cylinder damage, Engine misfire	Possible engine stop
Cylinder Head	Overheating	Fault Exhaust Valves	Cylinder damage, Smoke	Engine damage
Cylinder Liner	Leakage	Overheating	Cooling water in cylinder	Engine damage
Piston Crown	Hole in Piston Crown	Dripping of Fuel Valve	Escape of combustion gas crankcase	Engine performance reduction
Thrust Bearing	Improper Lubrication	Leakage	Thrust bearing damage	Engine damage

Table 4: Sample of Main Engine FMEA

4.4 ADDITIONAL CASE STUDY RESULTS

4.4 (a) Scenario *I*

This scenario aims to cluster data for the main engine case study. The input dataset represents actual engine data operating under normal operating conditions. After simulating the data in the trained SOM, the SOM clustered the data into clusters 49, 77 and 57.

As observed, the data have been clustered successfully in cluster 49 and 77 which represent parameters operating in normal conditions as represented in Table 3. Moreover, it can be noticed that a part of the dataset has been clustered in the SOM cluster 57, which has previously remained inactive during the SOM training results.

By using the Euclidean distance metric and the custom developed algorithm, the clusters with the shortest distance related to cluster 57 are displayed in Table 5 in order to potentially identify neighbouring clusters containing similar data.

Euclidean Distance	<0.1	<0.15	<0.2	<0.25	<0.3	<0.35
Similar Clusters	No	No	48, 56	No	No	42, 46, 77

Table 5: Similar clusters to cluster 57 based on Euclidean distance criteria

By examining Table 5 and in conjunction with Table 3 it can be observed that all clusters identified as neighbouring clusters to cluster 57 are inactive apart from cluster 77 which represents healthy engine performance data. Therefore, it can be stated that cluster

57 also represents healthy engine data. Thus, the input monitored data representing healthy data has been successfully clustered in the SOM in combination with the custom algorithm identifying clusters that have the shortest distance between containing similar data.

4.4 (b) Scenario *II*

In this scenario, the input data represents simulated data in which the fuel oil system monitored parameter, namely the fuel oil inlet pressure, is below the OEM recommended threshold of 6.5 kg/cm^2 as displayed in Figure 4. The SOM successfully clusters the data into cluster 68, which as observed from Table 3 represents data for which the engine fuel oil inlet pressure is below the set threshold.

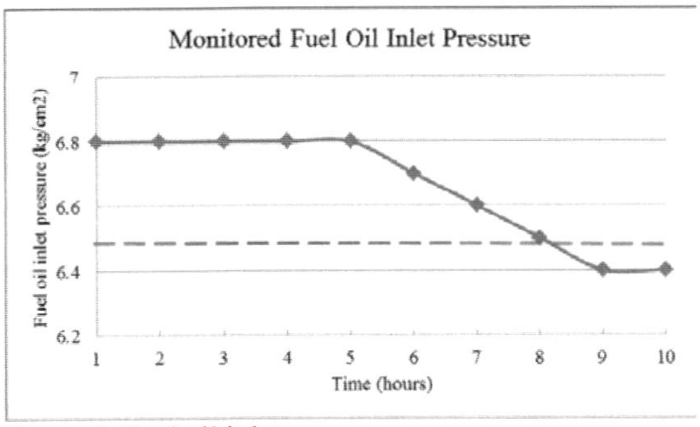

Figure 4: Fuel oil inlet pressure

As seen in Figure 4, the fuel oil inlet pressure remains constant at 6.8 kg/cm^2 for 5 hours and then drops and exceeds the threshold limit of 6.5 kg/cm^2 represented by the dashed line at the 8-hour mark. As this cluster represents data exceeding a predefined engine threshold limit, the developed in-house MATLAB code connects the associated cluster with the associated information contained in the main engine FMEA in order to present the results to the crew, ship operator or the onshore personnel. A sample of the relevant FMEA table for cluster 68, related to the fuel system is shown in Table 6 below.

Failed Item	Failure Event	Failure Cause	Local Effect	Global Effect
Fuel Oil Filter	Clogged	Contaminants Lack of maintenance	Fuel flow and pressure low	Engine speed drop
Fuel Pumps	Low pressure supply	Suction valve early or late operation	Erratic engine operation	Engine stop
Fuel Pumps	Abnormal sound	Bearing defective	Motor overloading	Engine output

Table 6: FMEA related to main engine fuel system

5. CONCLUSIONS

The marine industry is being reshaped by technology and over the next decade, new technologies will potentially transform shipping, an industry that is responsible for 90% of the global trade. As big data and the Internet of things is becoming a reality and the shipping industry is endeavouring to advance, there is no clear definition of big data and it is still challenging to quantify the volume of data required for successful machine learning, data training and analysis. This paper presented a novel approach for clustering data containing measurements of physical parameters for a ship main engine using the ANN-SOM methodology. The ability to better monitor performance and the data collected can assist to prevent future failures and business interruption. Shipping companies, owners and operators are taking a 'wait and see' approach when it comes to adopting smart shipping technologies, as key concerns regarding operating, staffing, insuring and implementing emerging technologies in the shipping sector are holding back the full embracement of the potentially game changing technologies. A key challenge for shipowners is measuring a baseline level of performance against which the output from any newly introduced system can be assessed and evaluated.

The unsupervised learning nature of the SOM provides a fast and efficient method to cluster data and model the underlying structure or distribution in the data. Moreover, through the Euclidean distance metric, data assigned in previously inactive SOM clusters, can be identified and labelled with near clusters sharing data similarities. The flexibility and adaptive nature of the developed SOM can be further expanded by examining the case in which input data are assigned in a cluster containing inactive neighbouring clusters. In this case, the cluster can be labelled accordingly by the user or ship operator for future reference and analysis. Furthermore, this implies that the SOM would also cluster similar data in the same cluster.

The trained SOM successfully clustered data in the case studies presented in Section 4. The SOM results were validated from actual data extracted from the main engine of a container ship. Additionally, the simulated data representing data exceeding alarm levels for the engine fuel oil system demonstrated the capabilities of the SOM clustering process in combination with the associated FMEA results. The SOM developed and presented in this paper provides a mechanism for processing data from ship sensors in order to further apply the processed data in a condition monitoring tool or software, utilising the processed data for prognostic, diagnostic and maintenance decision making activities. Relevant research can be found in Raptodimos et al. [24] and Lazakis et al. [25]. In addition, the SOM provides a diagnostic aspect, which connected with an FMEA can provide useful insight into possible engine faults and their causes, assisting decision making. Consequently, the diagnostic nature of the tool can be further expanded by also combining possible fault causes and remedies of the main engine and its subsystems and components. Similar methodologies and SOM algorithms can be developed for other shipboard systems and equipment.

ACKNOWLEDGEMENTS

The work in this paper is partially funded by Integrated Ship Energy & Maintenance Management System (ISEMMS) project. ISEMMS project has received research funding from Innovate UK under project No. 102435. The authors would also like to thank INCASS (grant agreement No. 605200) project for providing the data for this research. This publication reflects only the authors views and Innovate UK is not liable for any use that may be made of the information contained herein.

REFERENCES

1. Tester, K., et al., *Technology in shipping-The impact of technological change on the shipping industry*, in *Institute of Marine Engineers, Science & Technology (IMAREST)*. 2017: London, UK.
2. Raza, J. and J.P. Liyanage, *Application of intelligent technique to identify hidden abnormalities in a system: A case study from oil export pumps from an offshore oil production facility.* Journal of Quality in Maintenance Engineering, 2009. **15**(2): p. 221-235.
3. Tinsley, D., *Dawning of new era in asset maintenance*, in *Marine Power & Propulsion Supplement 2016*. 2016, The Royal Institution of Naval Architects: London. p. 4-5.
4. BS, *Condition monitoring and diagnostics of machines*, in *BS ISO 13372:2012*. 2012, BSI: London, UK. p. 28.
5. Jardine, A.K.S., D. Lin, and D. Banjevic, *A review on machinery diagnostics and prognostics implementing condition-based maintenance.* Mechanical Systems and Signal Processing, 2006. **20**(7): p. 1483-1510.
6. Kobbacy, K.A.H. and D.P. Murthy, *Complex system maintenance handbook*. 2008: Springer Science & Business Media.
7. Hagenauer, J. and M. Helbich, *Hierarchical self-organizing maps for clustering spatiotemporal data.* International Journal of Geographical Information Science, 2013. **27**(10): p. 2026-2042.
8. Kohonen, T., *Essentials of the self-organizing map.* Neural Networks, 2013. **37**: p. 52-65.
9. Principe, J.C., N.R. Euliano, and W.C. Lefebvre, *Neural and Adaptive Systems: Fundamentals through Simulations with CD-ROM*. 1999: John Wiley \& Sons, Inc. 672.
10. Pascual, D.G., *Artificial Intelligence Tools: Decision Support Systems in Condition Monitoring and Diagnosis*. 2015, USA: Crc Press.
11. Yan, J., *Machinery prognostics and prognosis oriented maintenance management*. 2014: John Wiley & Sons.
12. Kohonen, T., *The self-organizing map.* Neurocomputing, 1998. **21**(1–3): p. 1-6.
13. Isermann, R., *Fault-diagnosis systems: an introduction from fault detection to fault tolerance*. 2006, Berlin, Germany: Springer Science & Business Media.
14. Ben-Daya, M. and J. Knezevic, *Handbook of maintenance management and engineering*. Vol. 7. 2009, New York, USA: Springer.
15. ABS, *Guidance notes on Failure Mode and Effects Analysis (FMEA) for classification*, ABS, Editor. 2015: Houston, TX USA.
16. IACS, *Recommendation for the FMEA process for diesel engine control systems*, in *IACS Rec. 2014*. 2014: London, UK.
17. Cicek, K. and M. Celik, *Application of failure modes and effects analysis to main engine crankcase explosion failure on-board ship.* Safety Science, 2013. **51**(1): p. 6-10.
18. Jung, Y., et al., *A Decision Criterion for the Optimal Number of Clusters in Hierarchical Clustering.* Journal of Global Optimization, 2003. **25**(1): p. 91-111.
19. Turan, O., et al., *Investigating the reliability and criticality of the maintenance characteristics of a diving support vessel.* Quality and Reliability Engineering International, 2011. **27**(7): p. 931-946.
20. Mokashi, A.J., J. Wang, and A.K. Vermar, *A study of reliability-centred maintenance in maritime operations.* Marine Policy, 2002. **26**(5): p. 325-335.
21. Emovon, I., *Multi-criteria decision making support tools for maintenance of marine machinery systems*, in *School of Marine Science and Technology*. 2016, Newcastle University: Newcastle.
22. INCASS, *Deliverable D4.2 Stakeholders' data requirements*, in *INCASS-Inspection Capabilities for Enhanced Ship Safety*. 2014, EC FP7 Project: UK.
23. INCASS, *Deliverable D4.1 Specification of requirements for machinery and equipment*, in *INCASS-Inspection Capabilities for Enhanced Ship Safety*. 2014, EC FP7 Project: UK.
24. Raptodimos, Y. and I. Lazakis, *An artificial neural network approach for predicting the performance of ship machinery equipment*, in *Maritime Safety and Operations 2016 Conference Proceedings*. 2016. p. 95-101.
25. Lazakis, I., Y. Raptodimos, and T. Varelas, *Predicting ship machinery system condition through analytical reliability tools and artificial neural networks.* Ocean Engineering, 2017.

AUTHORS BIOGRAPHY

Mr Yiannis Raptodimos is a PhD candidate at the University of Strathclyde, Glasgow, UK. His research investigates the development of a hybrid machinery condition monitoring framework combining reliability tools with data-driven methods based on neural networks. He has industrial experience with regards to ship operations and maintenance and has participated and contributed in EU funded research related projects in the fields of maintenance and inspection, reliability analysis, data analytics, condition monitoring and decision support systems.

Dr Iraklis Lazakis is a senior lecturer in the Department of NAOME at the University of Strathclyde. He has 10 years of industrial experience with regards to ship operations involving maintenance and repair and surveys. He has developed his expertise on reliability and criticality based maintenance in order to improve ship safety and operations. He has participated and contributed in industry, UK and EU funded research related projects in the fields of maintenance, inspection, condition monitoring, decision support systems, reliability and risk analysis.

USER INTERFACE FOR BIG DATA ANALYTICS IN SHIPPING

Prabjot Singh Chopra and **G Umamaheshwar Reddy,** SeaPlus InfoNautics Pte Ltd, Singapore

SUMMARY

The Internet of Things (IoT) will spur the development of real-time data analytics to conduct real-time sensing, analysis, and response during ships operations. Innovations in storage and managed services have speed up the capture process. With large volumes of structured and unstructured data generated, smart analytical tools seamlessly connect to and combine a wide variety of cloud-hosted data sources, and intelligently process the captured data for meaningful analysis. The challenge now is in creating high intuitive, flexible, user interfaces that allow users to pull data from their traditional data stores as well as from unstructured data sources, then query and analyse using charts, images and graphs instead of by needing to write the endless lines of code required to work with the next generation data frameworks. This paper discusses the specifications for the user interface to be a platform with such an easy to use, intuitive interface, that totally operators could use directly.

NOMENCLATURE

UI	User Interface
GUI	Graphical User Interface
BI	Business Intelligence
GPU	Graphics Processing Unit
SSD	Solid State Drive
IoT	Internet of Things

1. INTRODUCTION

Industrial Internet has brought digitalization of data on-board of the vessel for ship operations. The evolution of IoT technologies is providing opportunities to exploit the explosion of digital data. Shipping industry is embracing the digital disruption to improve the performance, productivity and efficiency of ship operations. Other sectors in the industry, ship yards and designers are keen to open and adopt data-driven technologies. IoT technologies providing transparency through data processing and data analytics - bridging the gap between onshore and high seas with increased mutual support.

Big Data refers to the flood of digital data from many digital earth sources, including sensors, digitizers, and scanners, numerical modelling, mobile phones, Internet, videos, e-mails and social networks. The data types include texts, geometries, images, videos, sounds and combinations of each. The evolution of technologies and human understanding of the data have shifted data handling from the more traditional static mode to an accelerating data arena characterized by volume, velocity, variety, veracity and value.

Results of Data analytics need to be presented in a simple and intuitive way, so users can understand and use them to form effective statistics, analyses, predictions, and decision-making processes to be applied to production practices and business operations. For this reason the display technology of big data, as well as technology for interacting with data, plays an important role in the big data picture. This paper discusses user interface design considerations for visual analytics and compiles various data visualization tools for interactive front end.

2. USER INTERFACES FOR BIG DATA APPLICATIONS

The UI design, is closely related to the creation of user interaction, and is an essential component of data visualization in order to provide users with an interactive visual analysis capability. It gives leverage over traditional data collection and analysis using BI tools, following time consuming process, finally to issue static, rare view reports.

The key advantages of interactive data visualisation tools are:
- Enable real-time data analysis
- Support real-time creation of dynamic, interactive presentations and reports
- Governance dashboard that displays user activity and data lineage
- Allow end users to interact with data, often on mobile devices
- Allow users to share and collaborate securely
- Ability to visualize and explore data in-database as well as in-memory
- In-memory data compression to enable handling of large datasets without driving up hardware costs
- Touch optimization for use with touch-enabled mobile devices

Visualization becomes an increasingly attractive and feasible means for companies to effectively and efficiently access and explore data to gain new understandings and insights. However, there are several critical challenges to realize viable visualization solutions for shipping industry.

First, the size of a dataset can range from several Gigabytes to Petabytes, which becomes a major challenge. Without suitable system and data support, it is difficult to visualize and analysis these datasets. Second, advanced algorithms and computer architectures are required to support scalable data storage and retrieval such as GPU, SSD, and cloud computing. Third, environment datasets (for example, weather datasets) present diverse data types, which can be vector, raster, relation or binary. Examples include Shapefiles, images, tables, and LiDAR point clouds. Each data type can have its own unique data structures and require specific domain knowledge to process and interpret, incurring complex visualization and user interface operations.

2.1 APPLICATIONS IN SHIPPING

Typically, the data derived on-board of vessel - In the form of electronic information, derived through various sensors, and difficulties of capturing, storing, managing and analysing it due to its 5V characteristics, qualifies to be Big Data. These technologies are further adopted for Small Data applications – existing datasets with volume, variety, veracity and value for future improvements.

Identification of application users and their needs are essential for the development of user interface with suitable data visualization tools. Table 1 shows the typical interests of stakeholders in the industry.

Role	**Function**	**Big data application areas**
Ship operator	Operation	•Energy Efficiency in operation
		•Safe operation
		•Schedule management
	Fleet planning	•Fleet allocation
		•Service planning
		•Chartering
Ship owner	Technical management	•Safety operation
		•Condition monitoring & maintenance
		•Environmental regulation compliance
		•Hull & propeller cleaning
		•Ship Automation
Ship Yard	New building /Repair/ Modifications	•Yard Resource planning & management
		•Production planning and monitoring
		•Retrofit& modification
Ship designer	Design Spiral	•Design optimization •Automation

Table 1: Typical Big data applications areas in shipping.

Each application will have related parameters and data collected through sensors and data analytics result in specific data outputs obtained for data visualization.

Typical ship operations applications considers following inputs:

Voyage data

- Automatically collected data (IoT):

Date and Time	Rudder order and Response
Ship's Location	Engine & Thruster order and Response
Speed	Hull openings status
Heading	Watertight and fire door status
Bridge Audio	Accelerations and hull stresses
Communications Audio	Wind speed and direction
Radar	AIS
ECDIS	Rolling motion
Echo Sounder	Configuration Data
Main alarms	Electronic Logbook

- Noon report

Machinery Performance data
- Automatically collected data (IoT)
- Manual report data
- Maintenance data / trouble data

AIS data
- Satellite AIS / shore AIS(IoT)

Weather data
- Forecast / past statistics
- Anemometer / wave measurement (IoT)

Business data
- Cargo transport data

The user interface and data visualization tools to be deployed based on the application development platform and compatibility of supporting controls and tools for the backend databases, frameworks.

3. BIG DATA ANLALYTICS

Turning Big Data into Big Value, analytics is the process of analysing the data using statistical models, data mining techniques and computing technologies. It combines the traditional analysis techniques and mathematical models to derive information. Analytics and analysis performs the same function, however, analytics is the application of science to analysis. Big Data Analytics refers to a set of procedures and statistical models to extract the information from a large variety of data sets. Data analytics is an important part of Big Data Taxonomy as shown in figure 1.

A few major Big Data Analytics application areas are discussed below.

- *Text analytics:* The process of deriving information from text sources. The text sources forms of semi-structured data that include web materials, blogs and social media postings. Text mining in the area of sentiment analysis helps organizations uncover sentiments to improve their customer relationship management.

• *In Memory analytics:* In memory analytics is the process which ingests the large amounts of data from a variety of sources directly into the system memory for efficient query and calculation performance. In-memory analytics is an approach for querying data when it resides in a computer's random access memory, as opposed of querying data stored on physical disks. This results in vastly shortened query response times, allowing BI applications to support faster business decisions.

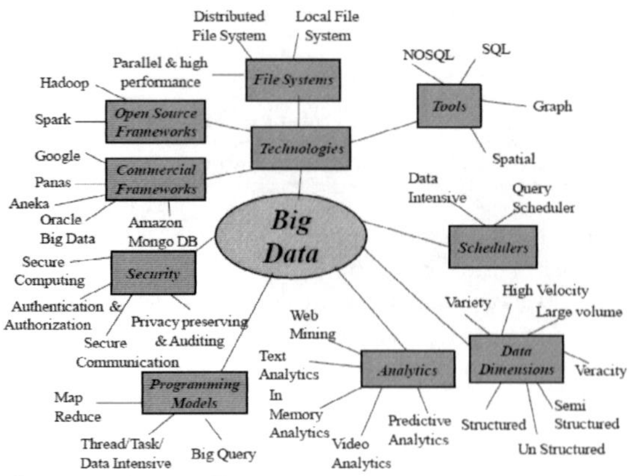

Figure 1: Big Data Taxonomy

• *Predictive analysis:* Predictive analysis is the process of predicting future or unknown events with the help of statistics, modeling, machine learning and data mining by analyzing current and historical facts.

• *Graph analytics:* Graph analytics studies the behavior analysis of various connected components, especially useful in social networking web sites to find the weak or strong groups.

Big data categories into two types-

- Operational Systems: For real-time, interactive work-loads where data is primarily captured and stored.

- Analytical Systems: Capabilities for retrospective, complex analysis that may touch most or all of the data points recorded. Each has their preferred tools and technology architecture.

Operational Systems tools:
Support unstructured databases (document databases like MongoDB). NoSQL databases are natively able to handle load by spreading data among many servers.

Analytical Systems tools:
MapReduce has emerged as the first choice for Big Data Analytics.

Combining operational and analytical technologies using Apache Hadoop, which has a storage part and a processing part.

4. USER INTERFACE SPECIFICATIONS

GUI, key feature of graph/visual analytics, is part of Human-Computer Interaction which is the study, planning and design of how people and computers work together. UI is what users see, hear, touch, talk to or control and direct. UI is often "the product" which users see and may make their decision on, which provides "user experience".

User Experience aspects to consider in UI Design:

- Application Features
- User Interface format
- Response Time
- Reliability
- Installability
- Help and docs
- Maintainability
- Graphics/attractiveness
- Be easy to use and intuitive

GUI is evaluated from user perception on usability:

- Stress, fatigue, panic, frustration, annoyance.
- Low productivity, errors causing increased cost of doing business

GUI – Design Principles:

Aesthetically pleasing	Familiarity
Clarity	Functionality
Compatibility	Forgiveness, recovery
Easy to understand	Predictability
Configurability	Responsiveness
Consistency	Simplicity
Control	Transparency
Context	Design trade-offs
Efficiency	

4.1 DATA VISUALIZATION

Data visualization is broadly defined as a method of encoding quantitative, relational, or spatial information into images. Various techniques or tools are used based on type of application and narration of information. Dashboards, charting and mapping are important tools applied in user interface design.

4.1 (a) Dashboards

Building an effective dashboard according to best practices for dashboard design is the culmination of a comprehensive BI process that would usually include gathering requirements, defining KPIs and creating a data model. However, the importance of proper dashboard design should not be understated – poorly designed dashboards could fail to convey useful information and insights and even make the data less comprehensible than it was originally.

Characteristics of good BI dashboard design is one that –

- *Makes the complex simple:* sensors provide lots of information, lots of data that changes all the time and different analytical needs and questions. Solutions are created taking all this complexity and make it simple.

- *Tells a clear story:* GUI to be able to connect data to its context in the business and to answer the user's questions. This is where the visual layout of a dashboard plays a crucial role.

- *Expresses the meaning of the data:* the chosen data visualizations need to correctly represent the data, and the information required to extract from it.

- *Reveals details as needed:* each user to have access to the data they need – no less but also no more. Some users might need to be able to see a more granular view of the data – others could suffice with an overview.

While each data dashboard has its own requirements, limitations, and goals, there are certain guidelines that are almost always relevant for dashboard creation. When designing a dashboard it's important to follow organizing principle. One of the most useful ones is the inverted pyramid. Display the most significant insights on the top part of the dashboard, trends in the middle, and granular details in the bottom as in figure 2. Each dashboard should contain no more than 5-9 visualizations. Select the appropriate type of data visualization according to its purpose.

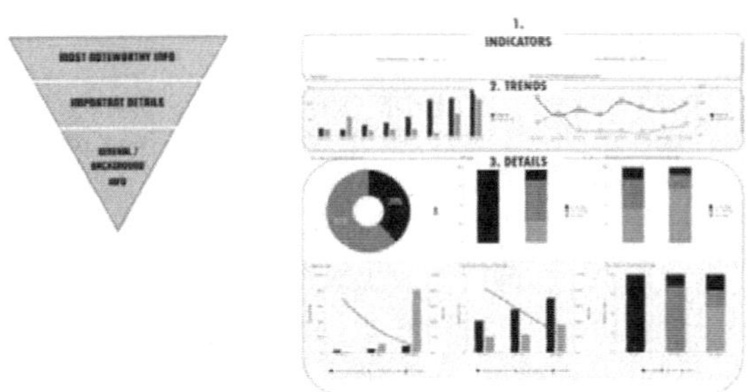

Figure 2: Dashboard Typical Layout

4.1 (b) Charting

Charting refers to the techniques used to communicate data or information by encoding it as visual objects (e.g., points, lines or bars) contained in graphics. The goal is to communicate information clearly and efficiently to users. Quantitative messages that users may attempt to understand or communicate from a set of data and the associated graphs, such as Time-series, Ranking, Part-to-whole, Deviation, Frequency distribution, Correlation, Nominal comparison.

Before choosing visualization for UI, consider which type of information been relayed as shown in figure 3:

- Relationship – connection between two or more variables.
- Comparison – compare two or more variables side by side.
- Composition – breaking data into separate components.
- Distribution – range and grouping of values within data.

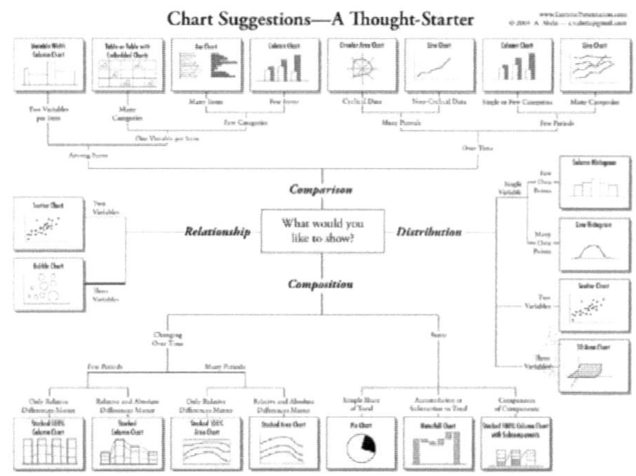

Figure 3: Charting methods

To perform these functionalities, charting methods are available such as, Line chart, Spline chart, Bar chart, column chart, Histogram chart, Pie chart, Doughnut chart, Area chart, Multi-axis chart, Bubble chart, Scatter Plot, Bullet Chart, Gauge chart, Combination chart and Grantt chart.

There are host of factors that may influence choice of a particular type of chart or visualization. Few factors are:

Size of Data points - this is probably the most important factor behind making a choice. For example, a Pie Chart may not work very well if the data point is more than say 30 but Line Chart may still be good choice. Similarly, an Area Chart or Line chart may represent a very large set of data.
Types of Data Points – check whether dealing with only one type of data or multiple different types of data that need to be plotted. Often combination charts are used if data types are different or different set of data need to represent. Example - using a combination of Bar Chart & Line chart.
Relationship of Data Points - If multiple Data Types are being presented in a Combination Chart, then if the different types of data are related or not also influence the choice of a Chart type.
Complexity Of Relationship - How the data sets are related matters in the selection of a chart. Complex relationships like many-to-many relations or multiple-relations between two data points are not easy represent

in regular Charts & Graphs and calls for creative visualization.

Static vs Animation vs Flow Representation - Most of the Charts and Visualizations really capture a snapshot or static state of data. In certain cases it may need to show a transition from one snapshot of data to the other (i.e. animated) or capture the states that the data pass through (i.e. data flow). A good example is Sankey Chart.

4.1 (c) Mapping

Geospatial data visualization of voyage planning, navigational route requires visual media in the form of maps and images, with textual description being an important subcomponent of such information. They are used for:

- Display quantitative information on a map.
- To present spatial relationships and patterns.
- When a regional context for the data is important.
- To get an overview of the distribution across geographic locations.

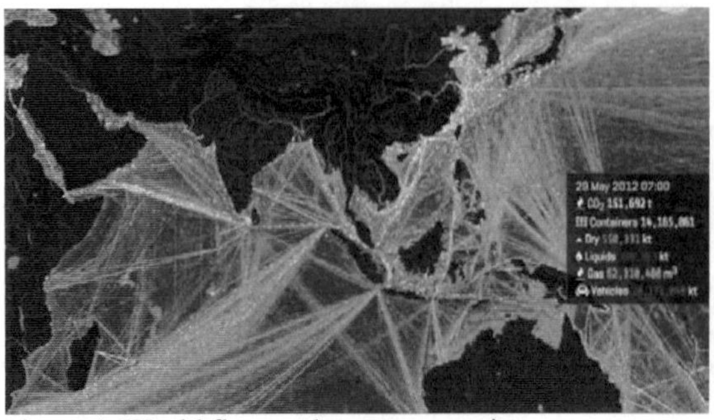

Figure 4: World fleet trade routes mapping

4.2 TOOLS FOR UI DEVELOPMENT

Under the vast scope of big data technologies variety of platforms are utilized to solve the big data challenges, where ready to use suitable visualization tool libraries are included. Though it is possible to develop custom made tools using Java, JSON, Jquery, .NET, R, Hadoop, MapReduce frameworks using proven libraries is most preferred method.

It is very essential to evaluate different to for the application needs by comparing following factors before selection of the tool:

- Price
- Power - to support large amounts of data
- Customization
- Data Migration
- Hosting
- Support
- Open Source
- Security
- Collaborative -allowing several people to work together
- Privacy
- Error-friendly
- Cross-platform -tool to work across different computer operating systems
- Mobile-friendly

4.2 (a) Dashboard Tools

Data Voyager:
Data Voyager is a data exploration tool that blends manual and automated chart specification. Voyager 2 combines PoleStar, a traditional chart specification tool inspired by Tableau and Polaris (research project that led to the birth of Tableau), with two partial chart specification interfaces: (1) *wildcards* let users specify multiple charts in parallel,(2) *related views* suggest visualizations relevant to the currently specified chart.

Tableau:
Tableau provides desktop, server and hosted software that allows users to connect, explore and visualize their data. Tableau has native connectors that enable it to query relational databases, cloud databases, flat files and spreadsheets.

Chartio:
Chartio is a cloud-based business intelligence solution that provides founders, business teams, data analysts and product teams in an organization tools to manage day-to-day business operations. Chartio's tools provide users functionalities for data exploration, enabling them to represent this data in form of charts. It also provides a central dashboard that displays specific information to the users.

Dundas BI:
Dundas BI, from Dundas Data Visualization, is a browser-based business intelligence and data visualization platform that includes integrated dashboards, reporting tools, and data analytics. It provides end users the ability to create interactive, customizable dashboards, build their own reports, run ad-hoc queries and analyse and drill-down into their data and performance metrics.
The software was designed to provide a high degree of customization and integration support through their programmable, open API platform, offering .NET, REST, and JavaScript APIs.

Power BI:
Power BI is a business analytics service provided by Microsoft. It provides interactive visualizations with self-service business intelligence capabilities, where end users can create reports and dashboards by themselves, without having to depend on information technology staff or database administrators. Dashboard is made simple with drag and drop gestures, and make decisions based on the right data anytime, anywhere, on any device. With Power BI dashboard is single view to monitor company's performance.

Quadrigram:
Quadrigram can create custom data visualizations in an intuitive way with the flexibility of a visual programming language. It enables prototype and share ideas rapidly, as well as produce compelling solutions with data in the forms of interactive visualizations, animations or dashboards.

Board:
Created to combine business intelligence, corporate performance management, and business analytics, BOARD is a full-featured business intelligence system that serves midsize and enterprise-level companies in a variety of different industry segments. Within the reporting functionality, BOARD allows users to pull from almost any data source, as well as generate full self-service reporting. The reports can be exported into several different formats, if necessary, such as CSV, HTML, and more.

InetSoft:
InetSoft Style Intelligence is a business intelligence software platform that allows users to create dashboards, Visual analyses and reports via a data mashup engine—a tool that integrates data in real time from multiple sources. These sources can include OLAP servers, ERP apps Web services, relational databases and spreadsheets.

Stats Panda:
Create Dashboard with Charts, Graphs & Data Tables. Convert static data to a Bar Chart or feed dynamic data to a Line Chart and share the insight with all stakeholders. Do statistical analysis on raw data and generate insight. Rich set of APIs to generate the Charts, Graphs & Datatables or do Statistical computation. APIs understand JSON input/output. Either make use of API Console or let Application call them using any popular programming language.

4.2 (b) Charting

D3-iD3:
iD3 is an open-source, cross-platform desktop application geared to simplify data visualization with D3 for non-JavaScript and JavaScript data scientists and analysts. With our integrated Python based data management system and custom generated attribute controls, iD3 supports the full process from importing data to exporting final presentation.

Vega-Lite:
Vega-Lite is a high-level grammar for interactive graphics. It provides a concise JSON syntax for supporting rapid generation of interactive multi-view visualizations to support analysis. Vega-Lite can serve as a declarative format for describing and creating data visualizations.

SVG:
SVG is Scalable Vector Graphics (SVG) is an XML-based vector image format for two-dimensional graphics with support for interactivity and animation. The SVG specification is an open standard developed by the World Wide Web Consortium (W3C). Widely supported. Traditionally, MS-Internet Explorer has been a laggard but starting with IE 8 SVG is supported in IE. Some of the popular libraries like morris.js, D3.js and others built on top of D3.js use SVG.

Canvas:
Canvas (HTML5) - Introduced as part of HTML 5 spec. example: flot, chartjs, jqPlot and other Javascript based visualization libraries.

Vega:
Vega is a visualization grammar, a declarative format for creating, saving and sharing visualization designs. Vega can describe data visualizations in a JSON format, and generate interactive views using either HTML5 Canvas or SVG.

Highcharts:
Highcharts is a charting library solution built using pure JavaScript, providing users with a quick and simple platform of integrating charts to their web sites or web-based applications. Highcharts offers a vast selection of supported chart types.

Ggplot2:
Ggplot2 is an R package for producing statistical, or data, graphics, but it is unlike most other graphics packages because it has a deep underlying grammar. This grammar, based on the Grammar of Graphics is composed of a set of independent components. ggplot2 can serve as a replacement for the base graphics in R and contains a number of defaults for web and print display of common scales.

Raw:
Raw is an open web app to create custom vector-based visualizations on top of the amazing D3.js library through a simple interface. Primarily conceived as a tool for designers and vis geeks, Raw allows to export visualizations in vector (SVG) or raster (PNG) format and embed them in web page.

Sankey Diagram Maker:
Sankey Diagram Maker is a new software tool to easily create Sankey diagrams, allow user to adapt Sankey flow charts to individual needs. It is the only tool that offers a free Sankey Diagram Maker.

Plotly:
Plotly offers a very engaging and user-friendly option for creating a range of visualisations within the browser. Only need to upload data, choose graph type, style it with

publication-quality layout and colour options, dive into the data to refine then publish and share.

Lyra:

Lyra is an interactive environment that enables custom visualization design without writing any code.

Google Charts:

Google Charts provides a perfect way to visualize data on website. From simple line charts to complex hierarchical tree maps, the chart gallery provides a large number of ready-to-use chart types. The most common way to use Google Charts is with simple JavaScript that embedded in web page.

Datamatic:

A new piece of browser-based software that runs on top of Google Sheets and allows for the easy, fast, code-free creation of a variety of both basic and complex interactive charts. It's highly recommended for multi-variate datasets on a deadline.

Dataseed:

A simple-to-use web app that allows uploading CSV files of complex, multivariate datasets and displaying them in a responsive dashboard of various visualization elements. It's great for exploratory data analysis.

Datawrapper:

A code-free, easy-to-learn copy and paste tool that creates a range of elegant, fully responsive interactive charts in seconds.

4.2 (c) Mapping Tools

GeoDa:

GeoDa is a free and open source software tool that serves as an introduction to spatial data analysis. It is designed to facilitate new insights from data analysis by exploring and modelling spatial patterns.

OpenData by ArcGIS:

The largest repository of geospatial and other civic datasets in the world, the Open Data platform has all kinds of handy data sources.

DataMaps:

Customizable SVG map visualizations for the web in a single Javascript file using D3.js

Google Maps API:

It is useful to build highly customisable maps with own content and imagery. It can create rich applications and stunning visualisations of data, leveraging the accuracy, comprehensiveness, and usability of Google Maps and a modern web platform that scales as data grow. Building and styling map can be done in only a few lines of JavaScript code.

Maptive:

Using this tool make sense of data on a Google Powered Map in seconds. Geo-analytics simplified. Spread sheet data can easily transferred into this mapping tool to provide great insights.

QGIS:

Quantum GIS (QGIS) is a user friendly Open Source Geographic Information System (GIS) licensed under the GNU General Public License. QGIS is an official project of the Open Source Geospatial Foundation (OSGeo). It runs on Linux, Unix, Mac OSX, Windows and Android and supports numerous vector, raster, and database formats and functionalities.

TimeMap.js

Timemap.js is a Javascript library that helps to produce online maps (such as Google, OpenLayers, and Bing). It can load one or more datasets (in JSON, KML, or GeoRSS) onto both a map and a timeline simultaneously. By default, only items in the visible range of the timeline are displayed on the map.

There are no perfect tools, just good tools for people with certain goals. Broadly in terms of analysis, flexibility and presentation, condition of the charting libraries and data visualization application tools are observed as in figure 5 (from Lisa Charlotte Rost, Dec 8,2016).

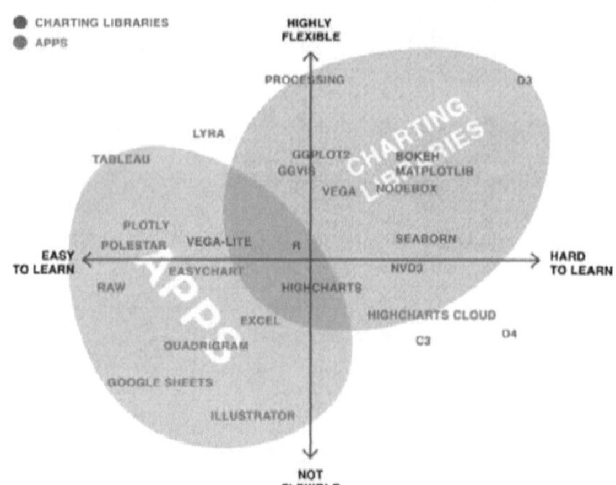

	STATIC	WEB - INTERACTIVE
APPS	ILLUSTRATOR, NODEBOX, EXCEL, POLESTAR, RAW	HIGHCHARTS CLOUD, QUADRIGRAM, EASYCHRT, DATAWRAPPER, TABLEAU, PLOTLY, GOOGLE SHEETS
CHARTING LIBRARIES	GGPLOT2, MATPLOTLIB, R, SEABORN, BOKEH, PROCESSING	D3, D4, C3, NVD3, GGVIS, HIGHCHARTS, SHINY, VEGA, VEGA-LITE

Figure 5: Observation of charting apps and libraries relative condition [5]

5. APPLICATIONS OF DATA VISUALIZATION IN SHIPPING

Vessel voyage monitoring dashboard:

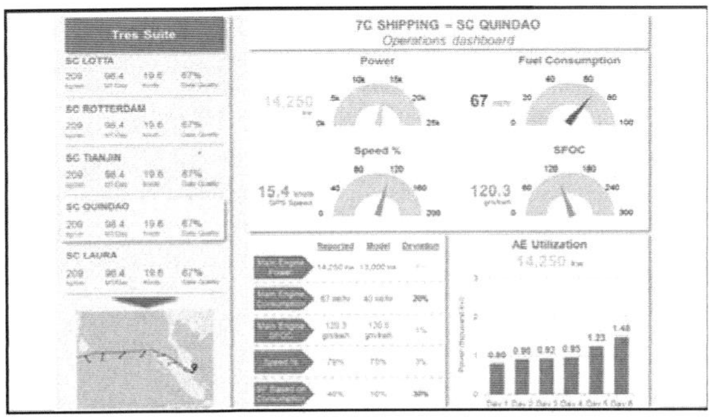

Operational optimization:
Visual analytics can be used to monitor and evaluate performance by identifying the trends and effects of variables such as trim, ballast, speed or time between maintenance. Weather routing takes advantage of weather and currents in order to optimize voyage distance or time travelled and thereby minimize voyage costs (or maximize profits) and maximize safety.

Maintenance alerts:
Hull and propeller performance monitoring acts as a decision support tool for determining condition assessment.

FOC optimization:

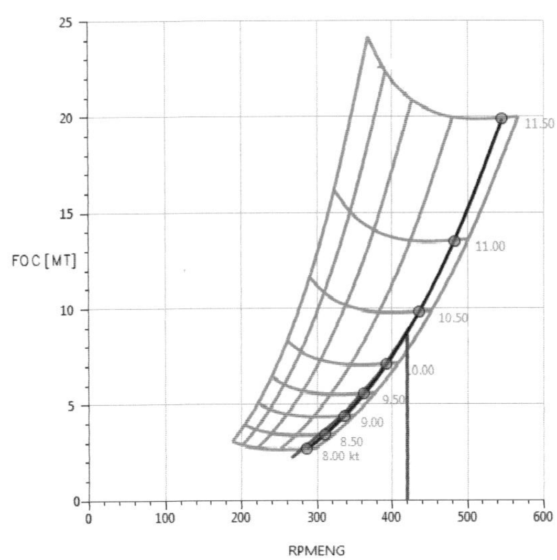

6. CONCLUSIONS

Big data visualization tools determine visualization tools could improve the understanding of ship operations, maintenance, and automation sensor data obtained by IoT. The clear, intuitive user interface of big data supports decision making by enabling onshore offices to quickly identify trends, develop a better understanding of the problem space, establish defensible baselines for monitoring activities, perform forecasting, and efficiency.

Big data visual analytics would provide a clear understanding of the relationships among all the variables, specifically those which cause increased costs and schedule overruns. Dashboards, Interactive charting, Mapping tools are very effective methods to be selected based on custom description in user perceptive.

Visualization tools are very effective for monitoring and evaluation of vessel performance, efficiency and Optimization.

7. ACKNOWLEDGEMENTS

We deeply thankful to Mr. Govinder Singh Chopra (Managing Director) and Mr. Dilip. G. Sarangdhar (Technical Director) of SeaTech Solutions International Pte Ltd for continuous support in monitoring and review of this work.

.

8. REFERENCES

1. Carl V. Lewis, Dataviz.tools: A curated guide to the best tools, resources and technologies for data visualization, Accessed at - http://dataviz.tools.

2. Ilan Hertz, ‘Dashboard Design Best Practices – 4 Key Principles’, Accessed at - https://www.sisense.com/blog/4-design-principles-creating-better-dashboards/.

3. Jānis Gulbis, ‘Data Visualization – How to Pick the Right Chart Type?’, Accessed at - https://eazybi.com/blog/data_visualization_and_chart_types/.

4. Satyajit Paul, 'Focal Point: 10 Things you must know about Charts, Graphs & Visualization', satyajit-paul.blogspot.sg/2015/10/10-things-you-must-know-about-charts-graphs-visualization.html.

5. Lisa Charlotte Rost, ‘What I Learned Recreating One Chart Using 24 Tools’, Accessed at - https://source.opennews.org/en-US/articles/what-i-learned-recreating-one-chart-using-24-tools/, December 8, 2016.

6. Hideyuki Ando, 'How to utilize Big data and IoTin the shipping sector ?', Japan - Norway Workshop, 12thFebruary 2016.

7. Hideyuki Ando, 'Performance Monitoring and Analysis for Operational Improvements', http://www.ship-efficiency.org/onTEAM/pdf/PPTAndo.pdf, International Conference on Ship Efficiency, 26-27 September 2011.
8. Raghavendra Kune, Pramod Kumar Konugurthi1, Arun Agarwal, Raghavendra Rao Chillarige, and Rajkumar Buyya, 'The Anatomy of Big Data Computing'.

9. AUTHORS BIOGRAPHY

Prabjot Singh Chopra holds the current position of Manager, Technical & Business Development at SeaTech Solutions International (s) Pte Ltd. He helms the Initial Design team at SeaTech Solutions, where he oversees the conceptualization for new designs and customers enquiries. His previous experience includes Structural Analysis of Jack-ups and jackets, Hydrodynamic Analysis on Offshore Fixed and Floating Structures, Earthquake Analysis using USFOS, Dynamic Analysis of Offshore Structures and Numerical Analysis of Free Vibrations including fully coupled interaction between structure and liquid.

Umamaheshwar Reddy Gadekari holds the current position of Assistant Manager at SeaTech Solutions International (s) Pte Ltd. He is responsible for ship design project management, marine software products development, design office project management & engineering applications development and performing design calculations. His previous experience includes managing design projects for tankers, OSV/DSV/AHTS, tugs, heavy lift & pipe lay vessels; software development for Ship Loadicator, Rule design calculations, Expert system for ship design software.

GAMING TECHNOLOGIES; ARE WE ON THE BRINK OF A NEW AGE OF HUMAN INTERACTION WITH NAVAL SHIPS?

N. Mitchell[a], **A. Anand**[b], **E. Gray**[c], **C. Shewell**[c] and **E. Trivyza**[c], UKNEST, United Kingdom
a. BAE Systems Submarines / b. BAE Systems Naval Ships / c. Babcock International

SUMMARY

Virtual Reality (VR) and Augmented Reality (AR) have been considered as a next step for the engineering world, particularly the psychology and technology behind gaming immersion, a trend which must be fully exploited in the naval industry.

This paper introduces the general fields of AR and VR, as well as the psychology and technology behind gaming immersion before refining to the naval sector. The initial surveys presented explore the current gaming technologies, providing an overview of the similarities and differences between the advances in the uptake of gaming technologies both in and out of the naval sector.

The future uses of AR and VR are explored, focussing mainly on ship design, and operator and maintainer interaction. Identification of key focus areas is important for the modern navy to adapt and take best advantage of the technology; with a particular focus on decision support, operator awareness and live-ship modelling.

The paper concludes with recommendations for the naval industry, in order to exploit the advancements in gaming technologies external to the naval sector to produce a smarter, more capable naval platform for the future.

1. INTRODUCTION

Technologies within Virtual Reality (VR) and Augmented Reality (AR) are becoming increasingly common in our everyday lives. The recent surge in the use of these technologies has lead to rapid enhancements in their capabilities.

The speed at which this technology has progressed is not only attributed to the 'user-pull' and demand of the technology, but also to the vast increase of computing power that means faster and more accurate graphics can now be achieved.

Previous Human-Computer Interaction (HCI) methods of the past forced the user to adapt to the limitations of the computer – greatly limiting the range of interaction methods possible. The initial HCI interfaces of switches and lights progressed to punch cards and paper tape, followed by the introduction of a computer mouse. Now, the standard method of interaction is centred on a touch-screen.

The developments in HCI have led to users opting to buy VR headsets in high street shops, to gain a more immersive user experience for both gaming and entertainment.

The vast developments in the commercial off-the-shelf (COTS) technology could have a significant impact in the military environment. This paper explores how these advancements should ultimately be adopted in the naval environment, specifically in the naval defence industry to create smarter, more capable defence platforms for the future.

1.1 VIRTUAL REALITY

VR is a computer-generated graphically represented environment that enables the user to experience a different reality to their present environment. The fundamental device within the VR system is a VR headset; the headset visually separates the users, from the space they are physically occupying.

The current age of virtual reality began in 2010, when Palmer Lucky created a first prototype of a VR headset, which has now evolved into the Oculus Rift VR headset.

The on-set of a variety of VR devices, ranging from Google® Cardboard to the HTC® Vive has allowed users, at every price-point to experience VR.

The concept of VR and its potential in the digital domain is widely recognised; with commercial technology beginning to mature due to advancements in the video-gaming sector. The naval sector can now benefit from the digital visualization tools represented by VR.

1.1 (a) Commercial Use of Virtual Reality

Entertainment – Applications such as Oculus® Cinema allow the user to become more deeply immersed in movie experiences; meanwhile applications such as LiveLike® VR allow the user to be immersed in a virtual experience of a sports match or concert.

Design – VR has been widely adopted in sectors such as architecture, to provide a fully immersive marketing environment before any construction occurs [1].

Gaming – VR within the gaming environment has now become commonplace to provide a fully immersive environment. As time progresses VR gaming is also now being applied to medicine for rehabilitation with patients who have suffered strokes [2].

Education – In recent years, a number of VR platforms have been designed and implemented to support education and training in a variety of learning domains. For example, the VR surgical training system Medical Realities, allows immersive training without endangering patients.

Sport – UK Sport have developed simulation VR technology that allows athletes to become familiarised with course layers ahead of major competitions by providing 360° and 3D simulations in a VR environment.

Space Training - NASA has been using VR technology for the past twenty years, primarily for familiarisation of the environment they are about to enter. NASA uses VR to train astronauts while they are still on Earth for experiences such as zero-gravity work environments and walking in space.

1.1 (b) Defence Applications of Virtual Reality

Platform Design – Ship design has benefited hugely from VR since design engineers are now able to identify, reduce and eventually eliminate errors more easily. Previous errors that occurred during the design phase, such as space and environmental constraints can be identified and thus transferring them to detailed design and production can be avoided. BAE Systems® Submarines have utilized VR technology within the design of the current class of submarines. Whilst not defence, Royal Caribbean® have implemented VR for ship design and 3D assessment of the Celebrity Edge cruise ship [3].

Real-Time Design Reviews – VR technology has been heavily used in the design of ships. BAE Systems® has used VR in the design of Type-26, [4] enabling real-time virtual design reviews – regardless of the location of the participating parties.

Platform Build – Virtalis® supplied BAE Systems® Submarines with VR enabled cabins and VR suites which were customised to create animations of major events in the boat's build. The VR cabins enabled the builders to view the 3D virtual models during build, in the cabins locally positioned to the boat, which provided a direct replacement for expensive physical models [5].

Training – BMT Defence Services has developed ENGAGE, a VR training platform that is designed to work across the full spectrum of military domains. ENGAGE can support both existing and new digital training scenarios due to its open and modular design. This in turn can present a more modern, flexible, and efficient training experience for military personnel currently and in the future. The ENGAGE training platform has been used to provide a 3D walk through tool for the Queen Elizabeth Class aircraft carriers [6].

1.2 AUGMENTED REALITY

AR is where the real-world environment is enhanced by computer-generated or extracted real-world, sensory input such as sound, video, graphics, or GPS data.

AR interaction experiences are different to traditional HCI, since the computer augments reality by providing computer-generated content and overlaying it on the real world. Because of this real-world interaction, the computer must adapt to what a human can understand, a role-reversal from previous interaction methods. Interaction methods based on speech, gestures, motion, and eye movement are more natural for humans, as these are the methods they routinely use to interact with the physical world.

AR is being implemented across a wide spectrum of industries and services. This section describes several examples of both commercial and defence applications.

1.2 (a) Commercial Use of Augmented Reality

The first commercial application of AR appeared in 2008, whereby a magazine advertisement of a model BMW® Mini, which when held in front of a computer's camera appeared on the screen. The user was then able to control the car on the screen to view different angles.

Augmented reality is being implemented across a wide spectrum of industries and services. This section describes several examples in non-naval applications.

Mobile Applications – The most common use of AR is for computer games, such as Pokémon Go®; which uses a smartphone's camera, gyroscope, clock, and GPS to enable a location-based AR gaming environment. Pokémon Go® users peaked at almost 45 million users in August 2016 [7]. This wide user base can be contributed to the level of accessibility of this AR application – users were using their own smartphone, in their own environment.

Healthcare – The first operation using Google® Glass was performed in 2013, whereby the primary benefit of AR is the ability to overlay vital information to a surgeon's field of vision during surgery [8]. Additionally, universities have started to adopt AR in the teaching environment with apps such as HoloAnatomy®; allowing insight into the biology of the human body. Though AR has not become main-stream in the medical domain, there are continuous on-going developments.

Data Visualisation – Systems such as Max Reality rely on AR technology to create dynamic data displays. Max Reality creates dynamic, 3D images of storms and atmospheric events, with the capability of integrating traffic data to inform of weather and traffic conditions along a route.

Visual Reconstruction – Mobile AR applications can allow people to 'step back in time'. Apps such as CityViewAR in Christchurch, NZ allow people to see how the city was before the numerous recent earthquakes and building demolitions.

Construction Sites – The Crossrail® contractor Laing O'Rouke® used the "In-Site" AR app , developed by Soluis® to transfer information about buildings or structures from The Cloud to the helmet and then overlay it as augmented reality on workers' visor screens. Following the successful use of Soluis' app a £1m government research grant has been awarded to develop an Augmented Worker Experience (AWE) for construction.

1.2 (b) Defence Applications of Augmented Reality

UK marine and defence companies such as BAE Systems®, Babcock Marine®, BMT Group® and Wärtsilä® have all invested in applications of AR, both within and external to the naval sector.

Tank Sighting Systems – The BAE Systems® Challenger® 2 tank features augmented reality in the sighting systems [9], representing a large step forward in both tank design and capability. Information overlays on both optical and thermal imager sights, allowing the Commander to have the data required within their vision.

Aircraft Cockpit – BAE Systems® plans to develop a new cockpit for combat aircraft, where by the dials and displays which feed information to the pilot are replaced with projected images directly in front of their eyes using helmets such as Striker II, which incorporates fully digital optronics and offers an enhanced situational awareness capability [10].

On-Board Maintenance – Virtual Service Engineer (VSE) concept by Wärtsilä® [11] allows the on-ship engineer to communicate with technical experts in Wärtsilä® remote service centres using specially designed goggles that utilise AR technology.

Additionally, Analytic Solutions business within Babcock® has developed an AR system to support maintenance activities on-board a submarine. The system overlays computer-aided design models of the real-life equipment, and other animations and videos, over the user's vision of the existing real equipment. In addition to providing an efficient way of viewing information, Babcock® are now exploring how the technology could improve communications in the support environment and dramatically shorten timelines for gaining permissions for critical engineering processes [12].

Operator Helmets – The US Navy is testing an AR Gunner Helmet "GunnAR" [13] which is designed to provide a Navy gunner with a Head-Up Display (HUD) of information directly from the vessel's gunnery liaison officer and weapon system in an easy-to-analyse visual medium. Additionally, the US Navy is also developing diver helmets to provide divers with real-time sonar, text messages, schematics, photographs, and videos [14].

In summary, the uptake of AR in the commercial environment has been vast, particularly due to the accessibility of this technology through COTS products, such as smartphones. The global AR market is expected to keep growing at pace to around $120bn in 2020 [15]. As highlighted, there has been uptake within the defence sector, with some key highlights in the naval industry. However there are key themes that could be learnt from both the commercial and other defence sectors to benefit the introduction of AR technology into sea-going vessels.

1.3 GAMIFICATION

The concept of 'gamification' is the application of game-design elements and game principles in non-game contexts, for instance where the principle of learning by 'click rewards' is introduced to the classroom environment.

1.3 (a) Gaming Interaction Methods

Gaming interaction psychology is a key part of 'gamification'. The gaming industry has been increasing the use of interaction psychology, leading to recognition of its potential in other industries.

Reward-Based Learning – the gaming industry 'teaches' users to interact with the game as the designers desire through the use of rewards. When a gamer interacts with the game in a specific manner the player will be rewarded, through gaining 'consumables' such as coins or ammunition. Reward-based learning has been employed in applications such as MyFitnessPal, where 'exergaming' encourages users to exercise more effectively and improve their overall health.

Additionally, IT companies such as Microsoft® use reward-based learning to train their users. In 2011 Microsoft® released the 'Ribbon Hero 2' game which educated users how to use the newly implemented 'ribbon' feature on Microsoft Office® 2010.

Multi-Display Systems - Immersion technology was used to assist in the design of public display systems, whereby using immersive technology the design of display screens could be maximised [16]. Video display walls, integrated by Quadvision® and Matrox® are now commonplace in visual advertising.

Customisable Displays – Many games such as World of Warcraft® [17] allow the user to personalise the information displayed within their field of view. Often this leads to the user placing low-variable, commonly referenced information near the centre of the display, with highly-variable information, such as alerts and warnings at the periphery.

Spamming – Many professional gamers also rely on the concept of spamming. One type of this is controller interaction spamming, whereby the user performs countless input interactions to the gaming controller, of which many are redundant. The countless interactions reduce the amount of effort increase during heightened 'action times'. Users often rely on this spamming technique to remain at a higher state of readiness, in preparation for intense battle scenes.

1.3 (b) Gaming Interaction Methods in Defence

Many studies on gaming interaction technology look at developing skills such as problem solving, critical thinking, communication, collaboration, and creativity [18] [19] through training and game-based learning.

Tactical Analysis - Games have been used in the military since the Roman Empire to create war scenarios for tactical analysis; these are often referred to as "War Games". Tables with icons representing units were used to create a visual representation of the battlefield, allowing the leaders to "play out" likely scenarios. This method became digitalised in the 1960s with early games such as the Naval War-Gaming System. These were effectively computer based mathematical models. Now there are programmes which can simulate the war time environment such as Modular Semi-Automated Forces (ModSAF) [20].

PTSD Treatment - The use of gaming technologies by the US military has three aims: recruit personnel, train personnel and treat of psychological disorders such as PTSD [21] [22]. Exposure therapy has been used by creating an immersive and interactive simulation of the environment which caused the illness.

Operator displays – BAE Systems® Submarines are investigating the use of customisable displays for operators, thus allowing the operators to personalise the data they are exposed to; allowing the operator to reach optimal performance through their personal settings.

In summary, it is noted that 'Gamification' is a concept that is becoming common-place in the IT and software environments. The uptake of gaming interaction and immersion techniques in the non-gaming environment is becoming more prevalent in every-day life. The naval industry appears to be falling behind other sectors, regarding the uptake of gaming techniques with only minimal developments using gamification techniques.

2 FUTURE GAMING TECHNOLOGIES IN THE NAVAL SECTOR

Following the analysis of VR, AR and gamification technique uptake in the naval sector it is clear that these gaming technologies have significant, unexploited potential in the naval environment.

Despite identifying areas where the naval industry has started to adopt these technologies; such as the use of VR in ship design and AR in ship maintenance there are other areas of the defence environment that have seemingly adopted these technologies at a faster rate.

Outlined below are both potential incremental applications and disruptive applications for the naval defence industry.

2.1 MAINTENANCE

Maintenance Training – Royal Navy maintenance training involves performing maintenance on exact replicas of ship's machinery within training rooms at HMS SULTAN. Although successful for training maintainers on replicas of the machinery, they however have little exposure to the environment surrounding the machinery on the ship. Training may benefit from being supplemented by real ship familiarisation using VR & AR to help induct trainees to ship machinery spaces prior to joining the vessel. AR could be used to overlay the ship's machinery space within the training room environment, thus delivering a more realistic learning environment.

Potential benefits of AR in maintenance training include:

- Reduced time for maintenance procedures;
- Reduced risk from potential issues that are identified during training;
- Increased "right first time" rate during maintenance operations;
- Improved safety during maintenance.

Maintenance Aids – With work at both Babcock® and Wärtsilä® paving the way for on-board maintenance it is imperative that the naval industry remains focussed on this approach. The increase in the amount of data and equipment on-board modern naval vessels can be counteracted using AR technical data overlays and virtual service engineers. Additionally, the ability for a maintainer to look at equipment and be able to access a full-service history could significantly reduce the time taken for on-board maintenance, compared to existing paper-based techniques.

Potential benefits include:

- Reduced downtime of critical naval assets;
- Reduced cost of re-work as maintenance tasks are more likely to be "right first time";
- Reduced number of specialist engineers required at any time.

Maintenance Period Planning – VR is already used widely in commercial ship design and maintenance, however is only just starting to be used in the naval sector.

Some of the potential benefits of the use of VR in maintenance periods are:

- Visualisation of the maintenance required;
- Accurately predict and minimise the amount of work required to allow for the maintenance of a system;
- Accurately predict and minimise the amount of work in wake that will be left following the maintenance of a system.

2.2 OPERATOR INTERACTION

Operator Displays – An increase in the amount of data obtained due to the increasing power and capability of ships sensors leads to a requirement for an updated method of interaction with this data. Using customisable displays, much like those of experienced gamers [17] could significantly enhance the levels of engagement and data absorbed by the users. Additionally, the concept of 'spamming' for heightened readiness levels could be beneficial for operators, particularly in submarines, where there is often significant downtime between periods of high activity.

Potential benefits include:

- Higher amounts of data available to operators;
- Higher performance levels at peak times;
- Less 'down-time' for operators.

Customisable Interfaces – The Royal Navy currently has a 3.5% deficit in recruitment [23]. To ensure that the UK maintains its naval power at sea it is imperative to discover ways of maintaining performance whilst reducing the man-power required. The use of customisable interfaces set up how the operator desires, as well as overlays of information would allow each operator to interact with more information. This higher level of interaction with information could reduce man power through combining roles such as the Navigation Officer and Helmsman.

Using gaming technologies has the potential to relieve the Royal Navy through:

- Combination of roles, meaning fewer personnel are required;
- Increased productivity/operator performance.

Technical Information Overlays – The potential to overlay critical information in an operator's field of view, through the use of COTS products, such as Google® Glass and the Smart Helmet by DAQRI® [24], has significant potential.

For ship's command, the bridge provides a location with clear visibility to the horizon, whereby the commander, Officer of the Watch, Helmsman and any other Able Seamen can safely navigate the ship, as well as efficiently providing commands to other parts of the ship via internal communications. VR and AR can enhance the capabilities and operational effectiveness of the Bridge in the near future.

The potential benefits of overlaying technical information into the operator's line of sight include:

- Instant access to data
- Information available without removing the real-world view, e.g. looking away from situation to a screen

Navigation Overlays – In 2005, the American submarine USS SAN FRANCISCO collided with an undersea mountain southeast of Guam [25]. The use of AR to overlay large amounts of information can significantly reduce the likelihood of accidents similar to this, particularly in submarine situations where there is no visual feedback. Through overlaying anti-collision plots, navigation charts and identified threats this enables the navigator to plot a desired course. Once plotted, the desired course could be followed like a navigation route on car satellite navigation systems.

Weather conditions are censored by receiving radio signals and by monitoring data from satellites on a visual display, whilst water depth is monitored using echo sound locators. Clarity when planning manoeuvres could be improved using VR. A VR headset could be used to simulate the weather conditions at a further position along the course and view the data that is received on-time, allowing personnel to visualise a 3D image.

The use of AR to provide overlays for navigational purposes has numerous potential benefits, which include:

- Reduced likelihood of ship damage due to adverse weather;
- Reduced likelihood of ship damage due to obstacles;
- Enhanced situational awareness.

2.3 PLATFORM MANAGEMENT

Multi-Asset Interaction – During operations where two vessels must come together, such as “Refuel at Sea” or the launch and recovery of off-board assets the use of VR and AR has high potential. The ability for the helmsmen of the two vessels to visualise the positions of both parties could significantly reduce not only the time-taken to bring two vessels together but also the probability of success in each approach.

With the on-set of the use of unmanned underwater vehicles (UUVs) in conjunction with submarines, the use of VR allowing the UUV operator could be visually immersed could significantly enhance the performance of both mission and recovery operations – as well as ensuring a human remains within the decision chain.

The potential benefits of using VR in multi asset interactions include:

- Remote operation – safer for the crew;
- Instant access to data for fast response;
- Reduction in probability of collision.

Emergency Scenarios - The systems and controls present on the bridge are numerous; therefore, through incorporating both AR and VR by introducing holograms, multiple operations could be undertaken with one centralised holographic system. Using technology such as the Microsoft® HoloLens means Radiation Hazard (RADHAZ) route plans for personnel could be viewed. A 3D ship hologram can be created, compartments can then be selected by touching the image and thus the best possible route can be calculated. Using voice recognition commands, much like the technology Amazon® Alexa uses [26], would further enhance the operator experience, thus reducing the need for man-power through having access to multiple stations on the bridges consoles.

The potential benefits of using AR in emergency situations include:

- Remote operation – safer for the crew;
- Enhanced situational awareness;
- Instant access to data.

Signature Management – With developments in sensor and detection systems in the naval environment it has become more crucial to improve the signature management of naval defence vessels; to maintain stealth. The use of AR to overlay signature information for machinery operators could be a breakthrough for ‘on-the-fly’ signature management, allowing the operators to have direct feedback on the effect of equipment on a ship’s signature.
The benefits of using AR to aid signature management include:

- Real-time signature management feedback;
- Operator awareness of equipment on ship’s signature;
- Reduction in ship’s signature.

3. CONCLUSIONS

This paper intended to identify areas in which gaming technologies could benefit the naval industry. Through identification of both commercial advancements in gaming technologies such as VR and AR it is evident that the naval industry should capitalise on the commercial developments.

In order to enhance the multi-billion pound assets that are built, maintained and operated within the naval defence industry there are two prospective options to introduce this technology; incremental and disruptive.

Incremental methods of introduction are likely to be more popular with operators, who prefer small changes. The potential for significant cost savings through the disruptive implementation of gaming technologies would be preferred by industry.

Many benefits would arise from the naval industry adopting gaming technologies as mainstream methods of data interaction and design toolsets. Man-power and processes can be reduced and simplified in all areas of naval vessels with the use of VR and AR. This not only reduces risks associated with harm or death to crew members, but also contributes to making vital steps towards creating smarter, more efficient and safer naval vessels for the future.

It is worth noting, the importance of maintaining naval operators’ link to the ‘real-world’. Use of VR, whereby the operator is fully immersed in a virtual world is not recommended and the importance of naval personnel being aware of their surrounding must not be underestimated. Therefore, it is perceived that AR technology is more suitable to the naval environment as this provides the user with both virtual information as well as awareness of the real-world situation.

In order to keep-pace with our every-day lives it is imperative that the naval industry exploits the advancements in gaming technologies external to the naval sector, in addition to adopting the principle of gamification to enhance naval vessels. Additionally, the naval personnel of the future are using VR, AR and other gaming technologies in their day-to-day lives; meaning the future designers and operators will be far more familiar and comfortable with this technology than the current operators and designers.

In summary, numerous areas where both incremental and disruptive uptakes of gaming technologies have been identified. The introduction of COTS technologies could significantly enhance the operator effectiveness, maintenance methods and platform management of naval vessels to produce a smarter, more capable naval platform for the future.

4. ACKNOWLEDGEMENTS

The authors would like to express their appreciation to Becky Thomas of BAE Systems Submarines for her valuable and constructive suggestions during the planning and development of this work.

5. REFERENCES

1. STOUFFS R , JANSSEN P., ROUDAVSKI S., TUNÇER B., 'What Is Happening To Virtual and Augmented Reality Applied To Architecture?' *Open Systems: Proceedings of the 18th International Conference on Computer-Aided Architectural Design Research in Asia*, (2013)

2. STOCKLEY R., O'CONNOR D., SMITH P., MOSS S., ALLSOP L. & EDGE W., 'A Mixed Methods Small Pilot Study to Describe the Effects of Upper Limb Training Using a Virtual Reality Gaming System in People with Chronic Stroke', *Rehabilitation Research and Practice*, (2017)

3. Royal Caribbean CEO Explains How Virtual Reality Is Changing the Cruise Industry, GAUDIOSI, http://www.alistdaily.com/strategy/royal-caribbean-ceo-explains-virtual-reality-changing-cruise-industry/ (Accessed: 8/11/2017)

4. Virtual Reality Technology Transforms Design of UK Warships, BAE SYSTEMS®, http://www.baesystems.com/en/article/virtual-reality-technology-transforms-design-of-uk-warships (Accessed: 28/11/2017)

5. BAE Systems® Submarine Solutions, VIRTALIS®, https://www.virtalis.com/blogs/casestudies/bae-systems-submarine-solutions-3/ (Accessed: 23/10/2017)

6. NEWELL, J., LUCK. S., 'Testing the boundaries of virtual reality within ship support', *EAAW VII Conference*, (2017)

7. Pokémon Go has 65 million monthly users, BUSINESS INSIDER, http://uk.businessinsider.com/pokemon-go-65-million-monthly-active-players-2017-4 (Accessed: 01/12/2017)

8. WEN R. et Al., 'Hand gesture guided robot-assisted surgery based on a direct augmented reality interface', *Computer Methods and Programs in Biomedicine*, (2014)

9. Challenger® 2, BAE SYSTEMS®, http://www.baesystems.com/en/feature/challenger-2 (Accessed on: 01/12/2017)

10. BAE SYSTEMS, 'Striker® II Brochure', *BAE Systems®*, (2016)

11. Virtual Engineering for Enhanced Services, WARTSILA, www.wartsila.com/resources/article/virtual-engineering-for-enhanced-services (Accessed: 23/10/2017)

12. Augmented Reality Seeing What Lies Beneath, https://defencescience.blog.gov.uk/2015/10/21/augmented-reality-adds-vision-to-defence-engineering-seeing-what-lies-beneath/ (Accessed: 26/10/2017)

13. Augmented Reality Gunner's Helmet (GunnAR), KELLER J., http://taskandpurpose.com/navy-augmented-reality-gunner-helmet, (Accessed: 12/11/2017)

14. US Navy Diver Augmented Vision Display (DAVD, BAGGALEY, K., https://www.popsci.com/navy-AR-diving-helmet (Accessed: 12/11/2017)

15. GOLDMAN SACHS, 'Virtual & Augmented Reality', *Equity Research*, (2016)

16. AZENGRUBER, B., HASLGRÜBLER, M., & FERSCHA, A., 'Engaging Interactions with Public Multi-Display Systems: Immersive Assessment', *HCI Lab*, (2015)

17. DYCK, J., PINELLE, D., BROWN, B. & GUTWIN, C., 'Learning from games: HCI design innovations in entertainment software', *Proceedings of Graphics Interface*, (2003)

18. QIAN, M., & CLARK, K. R., 'Game-based learning and 21st century skills: A review of recent research', *Computers in Human Behaviour*, (2016)

19. NOOR, A., 'Envisioning engineering education and practice in the coming intelligence convergence era - a complex adaptive systems approach', *Central European Journal of Engineering,* (2013)

20. SMITH, R. 'The long history of gaming in military training', *Simulation & Gaming*, (2010)

21. MEAD, C., 'War Play: Video Games and the Future of Armed Conflict', *Eamon Dolan/Houghton Mifflin Harcourt*, (2013)

22. Playing War: How the Military Uses Video Games, SHABAN, H., https://www.theatlantic.com/technology/archive/2013/10

/playing-war-how-the-military-uses-video-games/280486/ (Accessed: 12/11/2017)

23. MINISTRY OF DEFENCE, Royal Navy and Royal Marines Monthly Personnel Situation Report for 1 October 2017, (2017)

24. Smart Helmet, DAQRI, https://daqri.com/products/smart-helmet, (Accessed: 20/11/2017)

25. USS San Francisco Investigation Completed, AMERICA'S NAVY, http://www.navy.mil/submit/display.asp?story_id=18257 (Accessed: 02/12/2017)

26. Alexa Voice Service, AMAZON, https://developer.amazon.com/alexa-voice-service; (Accessed: 26/10/2017)

9. AUTHORS BIOGRAPHY

UKNEST provides a forum for the UK's professional naval engineering, science and technology community for addressing issues of common concern, fostering specific professional development needs, and giving a focal point for interaction with, and influencing the wider Government and industrial community.

FutureNEST is a collaborative team of graduates from across the UKNEST Member organisations. To produce this paper the FutureNEST members participated in workshops at member organisations as well as undertaking independent research.

Natalie Mitchell holds the current position of Systems Engineer at BAE Systems Submarines. She is responsible for work on future Combat Systems. Her previous experience includes roles as a Submarines Combat Systems Engineer and Technology Strategist within BAE Systems.

Alfred Anand holds the current position of Naval Architect at BAE Systems Naval Ships. He is responsible for initial bid design work within the Type31e Programme. His previous experience includes completing the BAE Systems Graduate Development Framework covering stability, testing & commissioning, finance, research & technology and structures.

Ewan Gray holds the current position of Naval Architect at Babcock International. His previous experience includes experience includes conceptual design and dockings.

Catherine Shewell holds the current position of Naval Architect at Babcock International. Her previous experience, obtained on the Babcock International graduate scheme includes concept design through to build programmes.

Eirini Trivyza holds the current position of Naval Architect at Babcock International. She is responsible for work on Concept Design Team Her previous experience includes involvement at different stages of vessel life cycle.

A RISK-INFORMED DECISION SUPPORT TOOL FOR HOLISTIC MANAGEMENT OF FATIGUE DESIGN, INSPECTION AND MAINTENANCE

G Zou and **K Banisoleiman,** Lloyd's Register Global Technology Centre, Southampton, UK
A González, University College Dublin, Dublin, Ireland

SUMMARY

Fatigue cracks threaten integrity of marine and offshore assets and need to be managed properly during the life cycles. However, the decision making process for fatigue design and maintenance are often disconnected and probably not be optimal with respect to life cycle total costs. This paper proposes a holistic decision support tool for jointly optimizing fatigue design, inspection and maintenance decision based on risk quantification and life cycle cost analysis, taking into account the uncertainties associated with fatigue deterioration, inspection performance and repair effect. The tool can be used to support risk-informed fatigue design; inspection and maintenance decision making, so that fracture risk associated with design and operation of marine assets are controlled with the minimum life cycle total costs.

NOMENCLATURE

N_F	Fatigue life (year)
$\overline{a_1}$	Fatigue strength coefficient ($N^4 \cdot mm^{-6}$)
$\overline{a_2}$	Fatigue strength coefficient ($N^4 \cdot mm^{-6}$)
m_1	Fatigue strength exponents
m_2	Fatigue strength exponents
$\Delta\sigma$	Stress range (MPa)
h	Weibull shape parameter (MPa)
q	Weibull scale parameter
D	Fatigue damage accumulation
n_i	Load cycles at the i stress range
N_{fi}	Fatigue capacity under the i stress range (year)
n_b	The number of stress range
N_0	Annual number of cycles
Γ	Complementary incomplete Gamma function
Υ	Incomplete Gamma function
S_1	Transition point of bi-linear S-N curve (MPa)
Δ	Fatigue damage at failure
da/dN	Crack growth rate
C	Parameter in Paris' equation ($N^{-4} \cdot mm^{5.5}$)
m	Exponent in Paris' equation
K_{mat}	Material fracture toughness ($N \cdot mm^{-3/2}$)
ΔK	Stress intensity factor range($N \cdot mm^{-3/2}$)
ΔK_{th}	Threshold or the stress intensity factor ($N \cdot mm^{-3/2}$)
$Y(a)$	Geometry function
a_0	Initial crack size (mm)
a_c	Critical crack size (mm)
a_d	Detectable crack size (mm)
T_{SL}	Service year (year)
Δt_{ins}	Inspection interval (year)
c_{d0}	Direct unit design cost
c_i	Direct unit inspection cost
c_r	Direct unit repair cost
c_f	Failure consequence
C_D	The costs of
C_M	The costs of maintenance
C_F	Failure risk
T_0	Reference plate thickness
T	Plate thickness

1. INTRODUCTION

Marine assets need to be designed and managed during their life cycles to fulfil serval requirements, e.g. safety, serviceability, reliability, robustness and sustainability. The challenges for safe operation of marine assets mainly come from harsh sea environments, in which fatigue cracking, corrosion and other mechanisms cause degradation of structural resistance. Compared with other mechanism, fatigue cracking and crack growth is more critical, as crack growth can result in sudden rupture of load-carrying components or even the whole structure. Moreover, due to the nature of crack evolution, the time for detecting cracks is often very short. As shown in Figure 1, fatigue cracks are typically very small for a significant part of service life and thus hard to be detected.

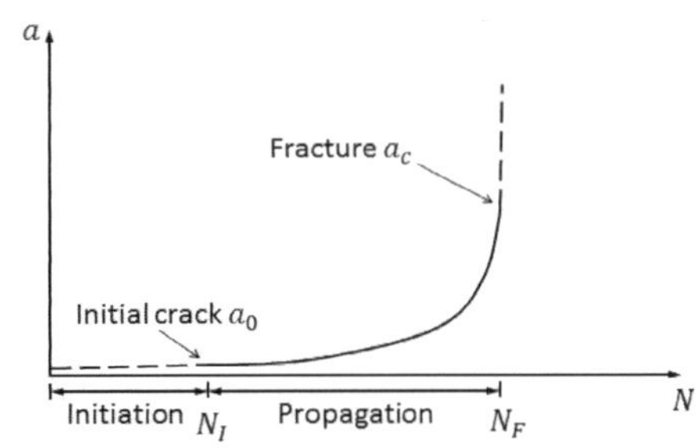

Figure 1: Typical crack development history

In order to meet the safety requirement, some measures need to be taken in life cycle stages of marine assets, as shown in Figure 2. Besides design fatigue assessment and manufacture quality control, operational inspections are essential for several reasons. First of all, it is typical not economical to design structural details in such a way that crack is unlikely to initiate in the lifetime of a marine asset. In addition, gross errors in design and manufacture can be found by operation inspection. What's more, damage or structural responses measured by inspections can be used to update failure probability as well as distributions of model parameters, and thus reduces uncertainties.

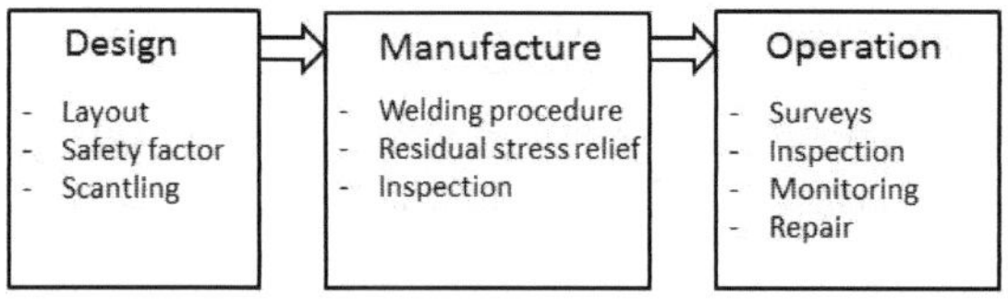

Figure 2: Safety control in the life cycles of marine assets

Fatigue deterioration process is associated with high degree of uncertainties, e.g. those associated with fatigue loading, calculation of load effect, material properties, test data on fatigue resistance, fatigue accumulation model and crack growth model, etc. There are also uncertainties in the performance of non-destructive testing methods and the effect of repairing method. Managing the uncertainties associated with design, inspection and maintenance decision making process with proper techniques and tools is a good start towards optimal decisions. Reliability and risk analysis methods have been applied in marine and offshore engineering to address the uncertainties and achieve optimal investments in inspection and maintenance [1-4]. However, the studies are generally on existing structures and fatigue design has seldom been connected with inspection and maintenance optimization.

The general idea of this paper is that assets are best manged with a holistic approach, jointly considering all engineering decisions (design, inspection and maintenance) and uncertainties affecting fatigue performance, based on the same experimental data for fatigue life. In this regard, a decision support tool is developed on the basis of life cycle risk and cost analysis. The tool can be used to support fatigue design, inspection and maintenance of marine assets.

2. PROBABILISTIC FATIGUE MODELING

Fatigue is the the phenomenon that cracks initiate and propagation till final fracture in structural details under cyclic loading that may be below material tension strength. Fatigue failure is caused by accumulation of damage cycle by cycle. Main factors contributing to fatigue failure are local stress concentrations, welding flaws and discontinuities, residual stresses, corrosion, etc.[5].

2.1 S-N APPROACH

2.1 (a) S-N Curve

S-N curves are developed to express the relationship between the fatigue life and nominal stress range based on statistical analysis of specimen test data. Each S-N curve represents a class of structural details. Equation 1 is formulation of typical two-segment S-N curves.

$$\begin{cases} N_F \Delta\sigma^{m_1} = \overline{a_1} & N_F \leq 10^7 \\ N_F \Delta\sigma^{m_2} = \overline{a_2} & N_F \geq 10^7 \end{cases} \tag{1}$$

where N_F is fatigue life, m_1 and m_2 are the fatigue strength exponents, and $\overline{a_1}$ and $\overline{a_2}$ are the fatigue strength coefficients.

2.1 (b) Fatigue Loading

Main source of fatigue loading is cyclic wave loading. The stochastics of wave loading can be determined by measurements or direct calculations together with statistical counting methods. It is generally acknowledged that two-parameter Weibull distribution as Equation 2 can be used to model the long-term stress distribution well.

$$F(\Delta\sigma) = 1 - \exp\left[-\left(\frac{\Delta\sigma}{q}\right)^h\right] \tag{2}$$

where F is cumulative probability function of $\Delta\sigma$, h is Weibull shape parameter and q is Weibull scale parameter.

2.1 (c) Accumulation of Fatigue Damage

Miner's rule is widely used as accumulation law for fatigue damage D, given by

$$D = \sum_{i=1}^{n_b} \frac{n_i}{N_{fi}} \tag{3}$$

where n_i is number of load cycles at the i stress range level; N_{fi} is the fatigue capacity under the i stress range level; and n_b is the number of stress range levels.

Employing Equation (1) - (3), the fatigue damage for two-segment S-N curve is formulated as follows,

$$D = N_0 \left[\frac{q^{m_1}}{\overline{a_1}} \Gamma\left(1 + \frac{m_1}{h}; \left(\frac{S_1}{q}\right)^h\right) + \frac{q^{m_2}}{\overline{a_2}} \Upsilon\left(1 + \frac{m_2}{h}; \left(\frac{S_1}{q}\right)^h\right)\right] \tag{4}$$

where N_0 is the annual number of fatigue loading, Γ is complementary incomplete Gamma function; Υ s incomplete Gamma function; S_1 is the transition stress range of two-segment S-N curve.

2.1 (d) Probabilistic Analysis

Limit state based on S-N approach is formulate as

$$h_1 = \Delta - D \tag{5}$$

where Δ is fatigue damage at failure.

Main sources of uncertainties associated with S-N approach are S-N data and Miner's rule. It is suggest the use of a bias of 1.0 and coefficient of variation (COV) of

0.3 for the modelling error caused by inaccuracies of the Miner's rule [6]. The uncertainties associated with S-N data are normally indicated in design codes and rules.

2.2 FRACTURE MECHANICS APPROACH

When it comes to inspection and maintenance planning, one has to know more details about the fatigue damage than the information provided by S-N curves and the damage should be measurable by inspection methods. Fracture mechanics (FM) represents a more rational tool for crack growth analysis, fatigue life prediction and inspection planning, as it is based on the physics of crack development.

FM formulations express the relationship between crack growth rate and the local stress field, which is the driving force for crack growth. Paris' law [7] first relates the crack propagation rate to the range of stress intensity factor, and is given by

$$\frac{da}{dN} = C\Delta K^m, \quad \Delta K_{th} \leq \Delta K \leq K_{mat} \tag{6}$$

where da/dN is crack propagation rate; C and m are material parameters; K_{mat} is material fracture toughness; ΔK is stress intensity factor range; ΔK_{th} is threshold value for the stress intensity factor range. The stress intensity factor range ΔK is given by

$$\Delta K = \Delta\sigma Y(a)\sqrt{\pi a} \tag{7}$$

where $Y(a)$ is geometry function and $\Delta\sigma$ is stress range.

Based on crack size, a limit state function as

$$h_2 = a_c - a \tag{8}$$

where a_c is the critical crack size.

One challenge for applying the FM approach is that the resulting crack growth prediction is very sensitive to some of input parameters, and statistical information for the variables usually not sufficient. Probabilistic methods have been recognised to provide a sound theoretical basis for treating both inherent variability and modelling uncertainty.

2.2 (a) Material Properties

Although influenced by the environment and loading condition, the material fracture property parameter C and m is generally seen as material properties. The uncertainties in C and m is believed to be originated from the inhomogeneities in material, measurement method, procedure and statistical method for parameter estimation[8]. In probabilistic analysis, C is typically treated as a variable and m is assumed to be constant.

2.2 (b) Stress Range and Geometry Function

The stress range $\Delta\sigma$ and the geometry function $Y(a)$ are very important parameters for crack growth and fatigue life prediction, as it enters into the formulation with the power of m. Detailed calculation of the stress range and geometry function by finite element modelling however is often time consuming. Some recommended geometry functions for common welded structural details can be found in [9].

2.2 (c) Threshold of Stress Intensity Factor

The threshold value of ΔK_{th} depends on many factors, e.g. the stress ratio, loading sequence, residual stresses, mean stress, etc. BS7910 [9] recommends using the value $63\text{N} \cdot \text{mm}^{-3/2}$ for welded joints. In this paper, the threshold is not considered. This is not supposed to induce much inaccuracy in prediction, because the effect of ΔK_{th} on crack growth prediction is small when a value less than $63\text{N} \cdot \text{mm}^{-3/2}$ is used [10]. In addition, the threshold is also not considered in S-N approach.

2.2 (d) Initial Crack/Flaw Size

A prerequisite for application of FM approach for fatigue analysis is that the sizes of initial cracks are known. However, it is not an easy task to obtain distribution of initial crack/flaw size for specific applications in engineering. The sizes of initial flaws depend on many uncontrollable factors in design and manufacture, e.g. materials, welding techniques, non-destructive methods, quality control procedure and human factors, etc. Effective measurements of initial flaws are hard to obtain due to sampling and measuring problems. A comprehensive review on the studies on the initial flaw size can be found at [11].

2.2 (e) Failure Criteria and the Critical Crack Size

As illustrated by Figure 1, the fatigue life and reliability results are not very sensitive to small change of the critical crack size, because crack develops very quickly at the final fracture stage. In engineering practice, critical crack size is often set to be equal to the component thickness, because the consequence of fracture failure typically negligible before through-thickness crack occurs. In this paper, the critical crack size is also set to be equal to the plate thickness.

2.3 CALIBRATION OF FM MODEL

FM approach provides sound theoretical basis for crack growth analysis, based on which inspection activities can then optimised with respect to the inspection times, methods, and costs. However, it is widely acknowledged that the modelling uncertainties associated with FM approach are greater than those associated with S-N approach, as S-N approach represent a more practical approach with plenty of test data. In this regard, it is proposed that FM model is calibrated to S-N curves. The benefits of calibration are that uncertainties in FM model are reduced, and that consistent safety and reliability level is achieved between using FM approach and S-N approach. There is however some difference in current studies with respect to the calibration criterion and calibration procedure. Methods and guidance are given by [12] on selecting calibration parameter and criterion while taking into account uncertainties associated with both FM and S-N approach. In this paper, calibrated FM model is used for calculation of initial failure probability.

3. QUALITY OF INSPECTION METHOD

There are detection and measurement uncertainties associated with inspection method. To address the uncertainties, the performance of inspection methods needs to be characterized with probabilistic terms before it can be properly integrated into the decision support tool for holistic fatigue management.

The detection capability of an inspection method is often characterized with the concept of probability of detection (PoD). PoD refers to the probability that an existing crack can be detected by an inspection method. The concept can be understood as follows. The smallest detectable crack size a_d characterizes the accuracy of any inspection method. However, there are inherent uncertainties associated with detecting an existing crack, so the parameter a_d is not a constant value. The parameter is modelled as a variable with a distribution. The probability of detection event can be formulated as

$$\mathrm{PoD}(a) = P(a > a_d) = P(a_d < a) = F_{a_d}(a) \quad (9)$$

where a is random crack size, and $F_{a_d}(x)$ is the cumulative probability function of a_d. From the above formulation, it can be seen easily that the PoD curve of an inspection method is equal to the cumulative probability function of the a_d.

4. REPAIR EFFECT

In the context of crack control, repairs are implemented to remove cracks, decrease crack size, stop or slow down crack growth in order to prolong fatigue life or decrease the failure probability within service life. In engineering practice, repair decisions on fatigue cracks are often made based on industry experience or expert judgement. The repair methods that are applied in marine and offshore engineering for fatigue cracks include

- Drilling a stop hole
- Welding
- Welding plus post-weld treatment
- Replacement
- Grinding

The repaired structural details by different repair methods generally show different damage states and fatigue performances. In the paper, it is assumed that the repaired structure has an initial crack size a_0^r, the distribution of which can be the same as that of the old structure or can be a different distribution.

5. ENGINEERING DECISIONS AGAINST FATIGUE

5.1 DESIGN DECISION

For fatigue design, fatigue design factor (FDF) is an important parameter that needs to be determined which indicates design fatigue safety reserve. Traditionally, operational inspection and maintenance is not a consideration in the fatigue design process. In this paper, the decision on FDF is an output of the decision support tool in which the FDF is optimized together with inspection and maintenance decision. The effects and benefits of planned inspection and maintenance activities on the structural performance and risk are assessed quantitatively before they are actually implemented. With explicit quantification of the benefits of afterwards inspection and maintenance, more rational decision making on the FDF can be achieved. The FDF determined by the holistic management tool is optimal with respect to life cycle total costs (LCC).

5.2 INSPECTION DECISION

Inspections are recognized as main means of identifying damage in structures. Operational inspections helps to validate structural design and identify new damage developed in service. A complete operational inspection plan specifies:

- Where to inspect?
- How many inspections in lifetime?
- When to inspect?
- How to inspect?

5.3 MAINTENANCE DECISION

As inspection results are unknown at the stage of design, planned inspection cannot give any useful information that can change the failure probability if no action is scheduled following inspection results. It is the repair activities that ultimately change the structural reliability. Following detection of cracks, decision makers have to decision on

- Whether to repair the detected cracks?
- When to repair the detected cracks?
- How to repair the detected cracks?

6. DECISION MAKING PROCESS

Design and management of marine assets requires continual decision making to identify and counteract the damage caused by cyclic wave loading so that the risk caused by the damage is mitigated and structural performance fulfils acceptance criteria. The engineering decisions optimized within this decision support tool are fatigue design, inspection and maintenance for marine assets. The tool is developed based on qualification of failure risk and life cycle total costs, taking into account the uncertainties in fatigue loading, structural resistance, inspection quality and repair effect. Figure 3 shows the framework of the decision support tool. Some essential aspects of the tool are describes as follows.

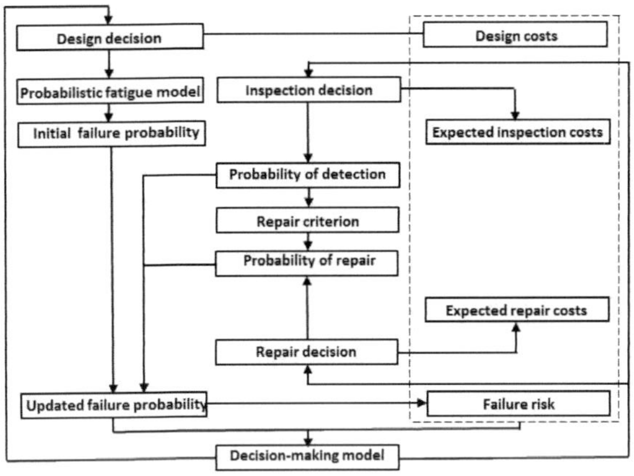

Figure 3: Decision support tool for holistic management of fatigue design, inspection and maintenance

6.1 UNCERTAINTY MODELLING

In order to make optimal decisions regarding fatigue design, inspection and maintenance, uncertainties affecting structural fatigue performance and failure risk should be identified. Parameters with high degree of uncertainties and have significant effects on the failure probability are modelled as variables.

6.2 OPTIMIZATION PARAMETERS

These are parameters that characterize the design, inspection and maintenance decisions, made to mitigate failure risk. In the meanwhile, measures set by the decisions typically involve heavy costs, and the net benefits that could be brought by those measures are dependent on the values of optimization parameters. These parameters are of great interest for decision makers and are outputs of the decision making model.

6.3 PERFORMANCE INDICATOR

It is an indicator based on what decision makers evaluate structural performance. The repair criterion is defined based on the performance indicator, and then optimised within decision support tool. Degradation of performance indicator under fatigue loading and the effect the planned inspection and maintenance on the performance indicator are explicitly assessed and formulated.

6.4 OPTIMIZATION OBJECTIVES

A basic objective for any engineering project is that the costs for the project are as low as possible. This paper addresses concerns about the whole service lives of marine assets, so an economic objective is that the life cycle total costs associated with design and operation of a marine asset are as low as possible. Requirement on the target failure probability can also be integrated.

6.5 CALCULATION OF PROBABILITIES

The initial failure probability is formulated with probabilistic FM approach. The planned inspections and associated repairs may decrease the failure probability. In order to formulate with updated failure probability, the probability of inspection and the probability of repair is formulated based on event tree analysis.

6.6 SOLVING STRATEGY

This refers to the strategy to find the optimization parameters time-efficiently. With the increase of the number of parameters, the complexity and difficulty in solving the problem increases, which necessitates introducing reasonable simplifications and some optimization techniques.

7. AN APPLICATION EXAMPLE

A typical fatigue-prone welded T joint is optimized with respect to the fatigue design, inspection and maintenance. For simplification, the same inspection method, magnetic particle inspection (MPI), is used in all inspections in the lifetime. The PoD is given by

$$PoD = F_{a_d}(a) = 1 - \exp(-a_d/m_d) \tag{10}$$

where m_d is the mean value of a_d. The mean value is about 0.89mm for MPI [1]. Other parameters and variables can be found in Table 1. The variables are material property C, initial damage size a_0, uncertainty in calculation of stress range B, and the detectable crack size a_d of MPI.

Two cases regarding the degree of uncertainty associated with $\log_{10} C$ and B are studied: low degree (LU) and high degree (HU) of uncertainties. The standard deviations (SD) in the LU case are shown in Table 1. The standard deviations are doubled in the HU case.

Cost structure is an input of the decision support tool. Four cost structures are applied in this example (CS1, CS2, CS3 and C4) and listed in Table 2 to represent different types of assets in terms of different cost ratios. The direct unit design cost c_d is assumed to be in proportional with relative plate thickness $\overline{T} = T - T_0$, i.e.

$$c_d = c_{d0} \cdot \overline{T} \tag{11}$$

The direct unit inspection and repair cost c_i and c_r are given as ratio to the cost of failure c_f.

In order to best understand the trade-off between investment in design for safety and in operational stage, equal time inspection interval is applied and the repair criterion is that detected cracks are repair immediately. The distribution of repaired crack size is the same as the distribution of initial crack size. The optimization parameters are FDF and inspection interval Δt_{ins}.

Parameter	Distribution	Unit	Mean	SD
$\log_{10} C$	Normal	$N^{-4}\cdot mm^{5.5}$	-12.74	0.11
m	Deterministic	-	3.0	-
B	Normal	-	1.0	0.15
N_0	Deterministic	cycle	5×10^6	-
$\log_{10} \overline{a_1}$	Deterministic	$N^4\cdot mm^{-6}$	11.855	-
$\log_{10} \overline{a_2}$	Deterministic	$N^4\cdot mm^{-6}$	15.091	-
m_1	Deterministic	-	3	-
m_2	Deterministic	-	5	-
T_{SL}	Deterministic	year	30	-
T_0	Deterministic	mm	25	-
a_0	Exponential	mm	0.04	0.04
a_d	Exponential	mm	0.89	0.89

Table 1: Input parameters and variables

Cost structure	c_{d0}/c_f	c_i/c_f	c_r/c_f
CS1	0.01	0.01	0.1
CS2	0.001	0.001	0.01
CS3	0.0001	0.0001	0.001
CS4	0.001	0.0001	0.001

Table 2: Four cost structures

8. RESULTS AND DISCUSSTIONS

8.1 OPTIMAL SOLUTION

It is found from Table 3 that with the increase of failure cost, the optimal solution is to increase the FDF and keep long inspection interval Δt_{ins} (with 2 inspections in lifetime) for the case of low degree of uncertainties in crack growth (LU) and low design cost. This is true even for the case $c_i/c_f = 0.0001$. However, with the cost of design increase one order of magnitude, the optimal solution is to decrease the FDF and decrease Δt_{ins} (increase the number of inspections). For the case of high degree of uncertainties (HU) and low design cost, the optimal solution is to increase the FDF and decrease Δt_{ins}. The optimal value of FDF decreases significantly with the cost of design increase one order of magnitude. High degree of uncertainties in the stochastic of crack growth generally requires bigger FDF and more frequent inspections. For some cost structures (CS1, CS4), only the FDF increases and the Δt_{ins} stays the same.

	CS1	CS2	CS3	CS4
LU	(3,10)	(5,10)	(9,10)	(1,3)
HU	(7,10)	(11,6)	(15,5)	(3,3)

Table 3: Optimal design and maintenance solution (FDF and Δt_{ins}) under different cost structure and degree of uncertainties in the stochastic of crack growth

8.2 THE MINIMUM LCC

From Table 4, it can be seen that the minimum life cycle total costs (MLCC) increase slightly with the increase of failure cost for the case of LU. This means that the optimal design and maintenance solutions are able to keep the MLCC and failure risks increase little with the significant increase of failure cost. For the case of HU, the MLCC is more difficult to control as indicated by Table 4. Table 5 lists the life cycle total costs corresponding to the cases in Table 4 but without any in-service inspection (WLCC). Table 6 gives the ratios of MLCC to WLCC. It is clearly shown that with the increase of failure cost, the ratio drops quickly, which implies with the increase of failure cost, the effectiveness and the benefits of the in-service inspections increase quickly.

	CS1	CS2	CS3	CS4
LU	11.26	12.33	13.86	14.70
HU	15.28	19.89	23.90	43.20

Table 4: The minimum life cycle total costs (MLCC) under different cost structure and degree of uncertainties in the stochastic of crack growth

	CS1	CS2	CS3	CS4
LU	21.87	95.22	239.10	6226.10
HU	23.50	134.18	932.80	3382.50

Table 5: Life cycle total costs corresponding to the cases in Table 4 but without in-service inspection (WLCC)

	CS1	CS2	CS3	CS4
LU	51.5%	12.9%	5.8%	0.2%
HU	65.0%	14.8%	2.6%	1.3%

Table 6: The ratio of MLCC to WLCC under different cost structure and degree of uncertainties in the stochastic of crack growth

8.3 RELIABILITY INDEX

It is shown by Table 7 that the reliability achieved by the optimal design and maintenance solution increases with the increase of failure cost, and decreases with the increase of design cost. By comparing the CS3 with CS4 in Table 3, 4 and 7, it is interesting to find that with the increase of design cost in CS4, the optimal FDF decreases, the optimal Δt_{ins} decreases (more inspections are scheduled), which leads to more LCC in CS4 than that in CS3, but on the contrary the reliability index achieved in CS4 is lower than that in CS3.

	CS1	CS2	CS3	CS4
LU	2.77	3.33	4.06	3.54
HU	1.98	2.98	3.44	2.83

Table 7: The reliability index β achieved by the optimal design and maintenance solution under different cost structure and degree of uncertainties in the stochastic of crack growth

8.4 SENSITIVITY ANALYSIS

It can be seen from Figure 4 that the reliability index β decreases with the increase of inspection interval Δt_{ins}. With respect to the life cycle total costs, it can be seen from Figure 6 that there is an optimal Δt_{ins}=6 which results in the minimum life cycle total costs $LCC_{min} = 19.89$. Generally, with the increase of inspection interval Δt_{ins}, the expected costs of maintenance C_M decrease, but the expected costs of failure C_F increase dramatically. However, if the inspection intervals lead to the same number of inspection, the expected costs of maintenance C_M for larger inspection interval is more that for smaller inspection interval. For example, for Δt_{ins}=6 and 7, the inspection number is both $N_{ins} = 4$; for Δt_{ins}=8 and 9, the inspection number is both $N_{ins} = 3$. The expected costs of maintenance C_M for $\Delta t_{ins} = 7$ are more than that for $\Delta t_{ins} = 6$. The same law applies to Δt_{ins}=8 and 9. This is because longer inspection interval means implement inspection later and the probability of detection and repair would be higher, so if the inspection number is the same, the expected costs of maintenance C_M would be more.

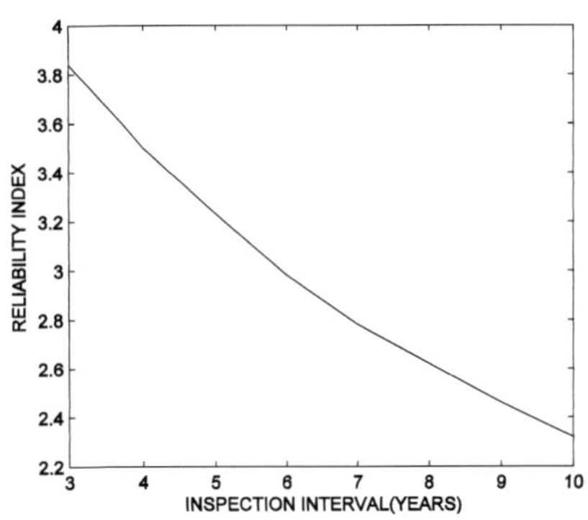

Figure 4: Sensitivity of the reliability index β with the inspection interval Δt_{ins} (HU, CS2, FDF=11)

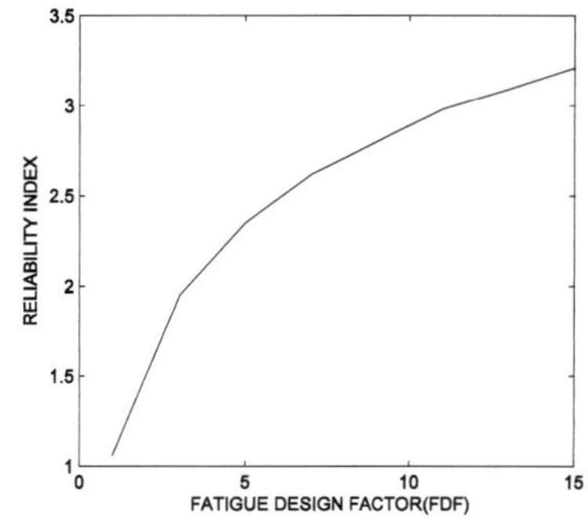

Figure 5: Sensitivity of the reliability index β with the fatigue design factor (HU, CS2, Δt_{ins}=6)

It can be seen from Figure 5 that the reliability index β increases with the increase of fatigue design factor (FDF). With respect to the life cycle total costs, it can be seen from Figure 7 that there is an optimal FDF=11 which results in the minimum life cycle total costs $LCC_{min} = 19.89$. With the increase of FDF, the costs of design C_D increase, the expected costs of maintenance C_M and the expected costs of failure C_F decrease. However, the costs of design, the expected costs of maintenance and failure and the LCC are generally not very sensitive to large FDF.

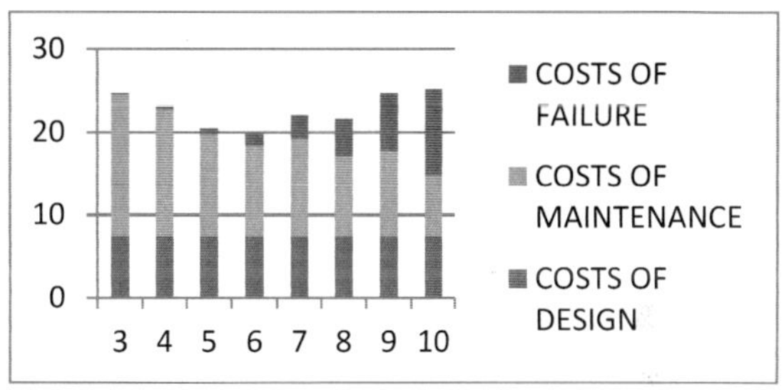

Figure 6: Sensitivity of the expected costs of design C_D, maintenance C_M, and failure C_F with the inspection interval Δt_{ins} (HU, CS2, FDF=11)

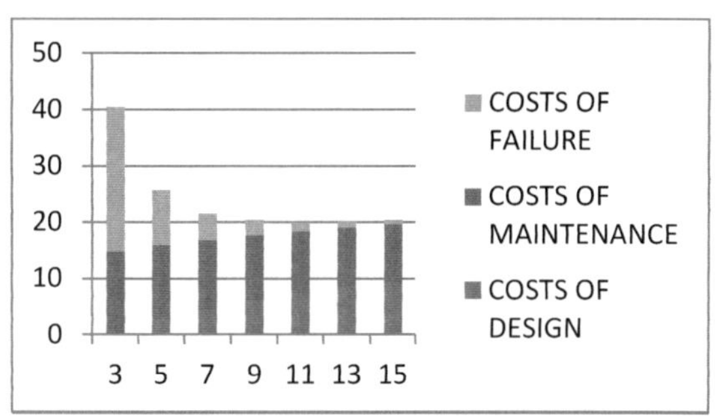

Figure 7: Sensitivity of the expected costs of design C_D, maintenance C_M, and failure C_F with the fatigue design factor FDF (HU, CS2, Δt_{ins}=6)

ACKNOWLEDGEMENTS

The authors would like to express their gratitude to the European Union's Horizon 2020 research and innovation programme for their funding toward this project under the Marie Sklodowska-Curie grant agreement No. 642453 (http://trussitn.eu).

REFERENCES

[1] Y. Dong and D. M. Frangopol, 'Incorporation of risk and updating in inspection of fatigue-sensitive details of ship structures', *International Journal of Fatigue*, vol. 82, pp. 676-688, 2016.

[2] M. Soliman, D. M. Frangopol, and A. Mondoro, 'A probabilistic approach for optimizing inspection, monitoring, and maintenance actions against fatigue of critical ship details', *Structural Safety*, vol. 60, pp. 91-101, 2016.

[3] S. Kim, M. Soliman, and D. M. Frangopol, 'Generalized Probabilistic Framework for Optimum Inspection and Maintenance Planning', *Journal of Structural Engineering*, vol. 139, no. 3, pp. 435-447, 2013.

[4] S. Kim and D. M. Frangopol, 'Optimum inspection planning for minimizing fatigue damage detection delay of ship hull structures', *International Journal of Fatigue*, vol. 33, no. 3, pp. 448-459, 2011.

[5] W. Fricke, 'Fatigue analysis of welded joints: state of development', *Marine Structures*, vol. 16, no. 3, pp. 185-200, 5// 2003.

[6] P. H. Wirsching and T. Y. Torng, 'Fatigue and Fracture Reliability and Maintainability Process', *Journal of Structural Engineering*, vol. 117, no. 12, pp. 3804-3822, 1991.

[7] P. Paris and F. Erdogan, 'A critical analysis of crack propagation laws', *Journal of Basic Engineering*, vol. 85, no. 4, pp. 528-533, 1963.

[8] T. Lassen and N. Recho, 'Fatigue Life Analyses of Welded Structures: Flaws', John Wiley & Sons, 2013.

[9] B. Standard, 'BS 7910 Guide to methods for assessing the acceptability of flaws in metallic structures', British Standards Institution, London, 2007.

[10] I. Lotsberg and M. M. Salama, 'Crack propagation versus crack initiation lives of FPSO weld details', *Proceedings of the ASME 2010 29th International Conference on Ocean, Offshore and Arctic Engineering*, 2010.

[11] G. Zou, K. Banisoleiman, and A. González, 'Methodologies for Crack Initiation in Welded Joints Applied to Inspection Planning', ICRSS 2016: 18th International Conference on Reliability and Structural Safety, Venice, Italy, 7-8 November 2016, 2016: World Academy of Science, Engineering and Technology.

[12] G. Zou, K. Banisoleiman, and A. González, 'Uncertainty quantification and calibration of a modified fracture mechanics model for reliability-based inspection planning', 12th International Conference on Structural Safety & Reliability (ICOSSAR 2017), Vienna, Austria, August, 2017, 2017: ICOSSAR.

AUTHORS BIOGRAPHY

Guang Zou is research fellow at Lloyd's Register of Shipping and PhD researcher at University College Dublin; funded by Horizon 2020 programme under the Marie Curie grant agreement No. 642453. He is responsible for asset management and life extension, probabilistic safety assessment, risk management. His previous experience includes research, design, modeling and testing of marine structures and underwater vehicles at CSSRC. He has been session chair of Fatigue Reliability in 36th OMAE and session chair of Fracture in 12th ICOSSAR.

Kian Banisoleiman is principal machinery consultant at Lloyd's Register EMEA. He obtained his PhD in the Department of Mechanical Engineering at University of Bath (1982). He is responsible for vibration analysis of rotating machinery, computer modelling of load distribution on contact lines of helical gears and gear tooth internal fatigue fracture analysis, thermal-mechanical simulation of turbocharged diesel engine performance, diesel engine health monitoring and fault diagnosis by performance monitoring, and field trials failure investigation analysis and application of finite element techniques to structural failure assessment. He has been Chair of the I.Mech.E.'s "Combustion Engines and Fuels Group Committee" and the "International Turbochargers and Turbocharging Conference Series".

Arturo González is associate professor at the College of Civil Engineering in University College Dublin (UCD). His research interests include structural dynamics, traffic loading; health monitoring, signal processing techniques and neural networks. He obtained his PhD from the School of Civil Engineering at Trinity College Dublin in 2000. In 2001 he continued research as Newman post-doctoral fellow in UCD. In 2003, he started his career as college lecturer in the School of Civil Engineering at UCD, where he is currently Associate Professor. In 2017, he received a Teaching Excellence Award from the College of Engineering and Architecture, and a Certificate recognising leadership at Ireland's Champions of EU Research 2017 from the National Support Network for Horizon 2020. He has been session chair and member of technical programme committee in many conferences.

THE PATH TO REAL WORLD AUTONOMY FOR AUTONOMOUS SURFACE VEHICLES

D Hook, CEng, MRINA, Senior Director – Business Development, ASV Global, UK,

SUMMARY

This paper presents new autonomy architecture, developed and tested over the past two years which has taken a "dumb" remote-controlled platform and built it into a state-of-the-art Autonomous Surface Vehicle, able to navigate safely at high speeds in a COLREG aware manner, across some of the busiest waterways in the world. The architecture is made up of a hierarchy of decision making components sharing a common world model. These components consist of: a route planner, path planner and last response engine. Each element of the hierarchy is working to reduce risk within decreasing time and space horizons, ultimately providing a minimum risk navigational solution. MAST and the collision avoidance system is now at the advanced prototype stage hence. Dstl and ASV welcome interest from other organisations who would like to explore and experiment with running their systems either on board or otherwise in command.

1. CONTEXT

ASV Global has built a reputation as the leading manufacturer of Autonomous Surface Vehicles (ASVs). In parallel to developing a wide product range of platforms, ASV Global has developed its own autonomous control system, ASView. ASView has been in development for 9 years and is actively used on over 80 platforms. It supports chart based operator situational awareness with pre-planned vehicle actions. The system currently relies on human input with limited on-board decision making.

To fuel growth of the industry, the goal moving forward is to develop a commercially useful autonomy system including and exceeding collision avoidance.

To ASV Global, real world autonomy means increasing the capability of the ASV to undertake decision making and thus enabling the remote operator to focus on safety critical issues and utilise natural human intuition. To provide the most efficient use of resources throughout the system as a whole, including the human element. In pursuit of this goal, ASV Global has developed an architecture which provides a robust foundation for the development of autonomy over the next decade. Throughout the design of the architecture a key element is the capability for expandability and future improvement, ensuring the system can meet current and future industry goals.

Initial tests and demonstrations of the system utilised data fused from radar, cameras and AIS sensors. This has been used in proof of concept demonstrations and showcased at the recent Unmanned Warrior naval exercise in the UK, deployed on multiple ASVs. Collision avoidance behaviours, taking into account important aspects of the collision regulations at sea, have been a critical part of this early development around which operational behaviours have been built.

These developments are accelerating the adoption of ASVs in the maritime industry by enabling capabilities such as over the horizon operations and the control of fleets of ASVs with minimal human input. Examples of applications spread across both military and commercial domains and include ISTAR, hydrographic survey, and support of underwater systems and in the future, commercial shipping.

The programme was directly funded by UK MoD through Dstl and led by ASV Global in partnership with Roke Manor Research and Cambridge Pixel.

2. PROJECT INTRODUCTION

The developed capability centred around the guiding vision of a high speed (>30knots) Unmanned Surface Vehicle (USV), notionally based outside Portsmouth Harbour, that could be tasked to autonomously intercept and investigate a target many nautical miles out at sea. This required development of high speed collision avoidance algorithms and on the fly route, path and trajectory re-planning all running directly on-board the USV for total self-sufficiency. The USV used was the Dstl Maritime Autonomy Surface Testbed (MAST) vehicle based on a bladerunner racing hull.

Figure 1: The MAST platform is a 34ft Bladerunner racing hull retro-fitted with ASV control system and extensive additional testbed hardware and software

The programme focused on researching what could be achieved using only current COTS sensor systems including AIS, radar, cameras and accelerometers which was successful and demonstrated that a great deal can be achieved using this approach. However, sensing remains a crucial performance driver and that there are many system/platform level trades-offs to be considered and no obvious "one-size-fits-all" sensor suite.

The concept and architecture was developed from scratch and is based on directly manipulating risk and uncertainty. This gives two key benefits: firstly a direct way for users to dynamically adjust risk appetites as mission goals change; and secondly a continuous real-time visualisation of "what I am thinking" over the next few seconds, minutes and hours for intuitive operator supervision and trust building. This contrasts with many collisions avoidance systems that are purely reactive, and hence give no advanced warning about their intentions.

The system was incrementally developed with an agile customer demonstration drum beat and numerous R&D testing activities including 54 days of MAST on the water R&D and testing time. This practical real-world approach is significant as handling the real world noise was consistently more challenging than expected in all areas, and rapidly raised the TRL of the subsystems. As a result the programme has delivered hands-on capability and created the foundations of a testbed that is available for further exploitation.

A competitor analysis carried out late in the programme showed that this is potentially the fastest full autonomy USV capability in the world; that the MAST platform creates more sensor processing challenges than most competitors; that other collision avoidance systems are more reactive and tend to plan over much shorter timeframes; and has a sensor fit that is broadly similar/comparable to other approaches. In short, this programme is one of the world leaders in this field to the best of our knowledge. Equally this is evidenced by the fact ASV Global have licenced the technology developed back from Dstl via the Ploughshares organisation, for a significant upfront cost and potentially recurring license fees as it is commercialised.

A further indirect benefit is that this programme has been instrumental in developing better working relationships with local and national authorities including Southampton Vessel Traffic Service, Portsmouth Queens Harbour Master and the Maritime Coastguard Agency. These authorities approved autonomous operations for this programme in their busy controlled waters, something that would have been highly unlikely just 2 years ago.

3. TECHNOLOGY OVERVIEW

Much of the development work undertaken in this programme is commercially sensitive and hence much of the system detail is deliberately restricted to high level discussion.

3.1 SYSTEM OVERVIEW

The Advanced USV System runs directly on board the USV and consists of a hierarchy of decision making components sharing a common world model. Each element of the hierarchy works to reduce risk with decreasing time and space horizons, ultimately providing a minimum risk navigation solution:

- The *Route Planner* generates routes over long distances taking account of water depth (and

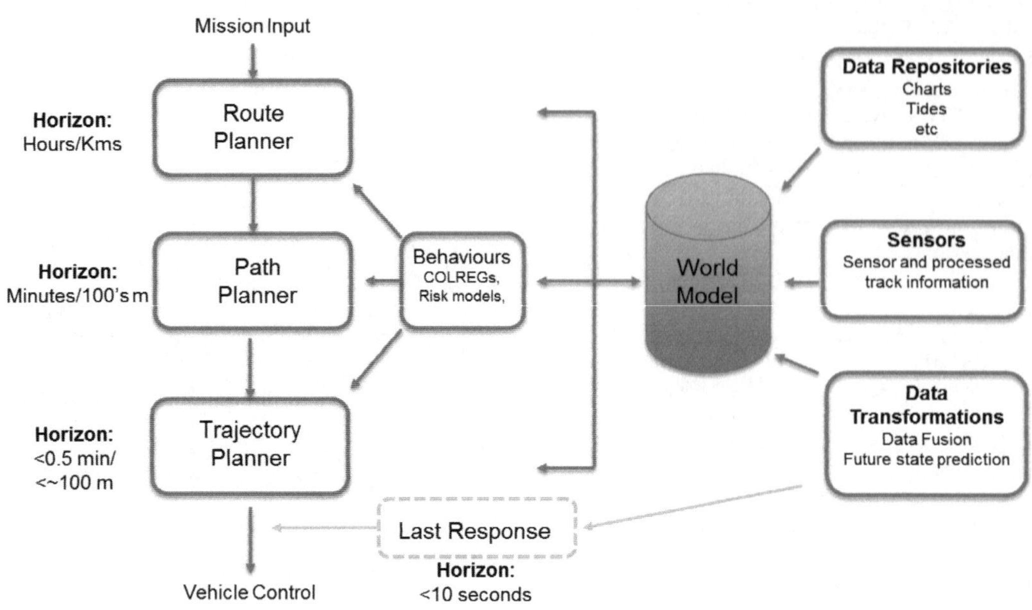

Figure 2: Schematic overview of the Advanced USV system

potentially shipping lanes).
- The *Path Planner* considers the next few minutes, choosing paths that minimises collision risk in a COLREG compliant fashion with other vessels.
- The *Trajectory Planner* manages the next few seconds to smoothly control and predict USV motion based on a dynamics model.
- The *Last Response* monitors the immediate vicinity and provides a safety critical engine cut out, if an unexpected obstacle is detected.

The system includes a sensor processing pipeline, with a 360° panoramic camera, broadband radar, AIS sensor, compass, GPS and an IMU, all of which is combined into tracks and vessel position information in a world model. Further processing then calculates track statistics and generates future trajectory predictions for targets. These predictions are processed based on COLREG rules within a 3D risk landscape to calculate risks of being in the wrong place at the wrong time. Chart features such as buoyage and shore data is also added to the risk landscape for use by the path planner.

The system is at the point where it would be possible to command MAST to autonomously navigate around the Isle of Wight under suitable weather conditions.

3.2 PROGRAMME ACHIEVEMENTS

The Advanced USV programme was predicated on a series of demonstrations to showcase the development of advanced autonomous navigation.

- **Demo 1 (November 2015)** demonstrated a basic risk-based path planner undertaking simple collision avoidance in an open water environment at low speed (~15 knots) using only AIS sensor input from a single target. Achieved capability for one-on-one crossing situations.
- **Demo 2 (March 2016)** added collision avoidance with multiple vessels using a fused AIS and radar sensor input and sea state monitoring to limit speed. Achieved capability for reasoning about compound multi party crossing situations.
- **Unmanned Warrior 2016 (October 2016)** The MAST and collision avoidance system was used extensively in military exercises over a three week period around the Isle of Skye. Achieved long endurance testing in more hostile environments with additional behavioural modes of operation.
- **Demo 3 (November 2016)** demonstrated multiple vessel collision avoidance at higher speed with automated last response functionality and basic route planning. Achieved the ability to operate fully unmanned at 30knots with autonomous emergency stop.
- **Demo 4 (February 2017)** demonstrated the addition of an initial vision processing system, a trajectory planner, AIS metadata processing to support detection/classification of objects and live extraction of chart data (both depth and objects). Achieved the integration of vision processing detections, basic vessel classification and dynamics aware planning for more advanced COLREG compliance.
- **Demo 5 (March 2017)** Added improved vision processing and showcased an integrated multi-phase "representative mission" on a long transit through shipping lanes and controlled waters. Achieved an end-to-end mission with "unscripted" collision avoidance. (Note: severe weather on the day unfortunately cut short the demo on safety grounds, but the system operated entirely successfully as expected).

3.3 SPECIFIC TECHNOLOGIES DEVELOPED

3.3 (a) Core System

One of the key insights from this programme was the representation of rules within the International Collision Regulations at Sea (COLREGs) as dynamic mathematical functions to represent risk. The system uses information about the relative positions & velocities of obstacles and the USV to classify it, and then generate risk distributions based on the predicted position, appropriate avoiding action and the error/uncertainty.

With a heavy emphasis on real world trials, the programme demonstrated that there are large numbers of challenges in handling the real world noise. Systems and algorithm that seem promising and successful in simulation in the laboratory had difficulty on the water. This level or ruggedisation challenge should not be underestimated. To give one example consider a basic COLREG – "overtaking vessel must keep clear". In human terms this is easily understandable, but for a machine we need to define when an overtake begins and ends, precise angles, relative velocities and thresholds. In theory this manageable but now what if you find yourself on the threshold one of these – how much hysteresis or fuzzy logic is needed? What if there is error and jitter in both detection and control? What if you loose and reacquire (an apparently different) vessel track midway through but need to remember that you are overtaking? What if you or it alters course and how is that differentiable from noise? What are the processing latencies in the system and how do you account for environmental effects and vessel dynamics? How do you ensure safety at all times? Some of these issues can be explored in simulation, but it is difficult to fully simulate the true level of real world uncertainty and plethora of special cases without extensive on the water experimentation.

3.3 (b) Sensing Sub Systems

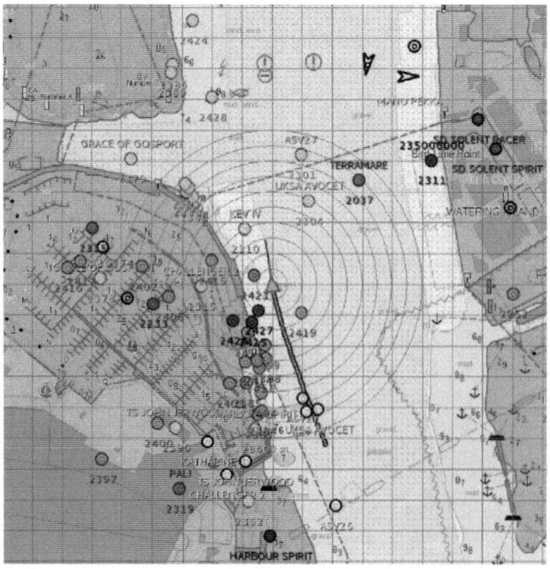

Figure 3: An example of AIS, radar & vision sensor outputs all being fused into tracks

MAST was fitted with a SIMRAD 4G broadband radar which is the most critical sensor of the collision avoidance system. Configuring an controlling the radar, plot and track extractor on just a small mobile platform was an ongoing an persistent challenge. A significant amount of effort was spent working with the supplier of the radar track extraction system, in order to optimise the performance in a variety of conditions.

The 360° panoramic Pt Grey LadyBug 5 camera was used for visual object detection. Work was undertaken to explore various detection algorithms including: Image masking, Horizon detection, Colour detector, Edge detector, Local Variance detector and particle probability hypothesis density (PPHD) filter.

Other sensor related work undertaken included the numerous "enabling sensors" such as compasses & GPS which again although seem conceptually simple had many challenges. Monitoring the vertical acceleration of the boat using IMUs with FFT & physics based algorithms to infer maximum safe speed was also developed.

Database algorithms were also developed to read and process electronic chart information for extraction of information relevant to collision avoidance for example water depth information, tide data and chart features such as buoys and marks.

3.3 (c) Final Outcomes and Deliverables

An internal comparison with other research activities in the field was undertaken in December 2016. This re-affirmed our belief that the work undertaken in this programme is at the forefront of the ongoing research activities in the industry and highlighted:

- The capability developed under this project is potentially the fastest full autonomy capability in the world.
- The light displacement, high speed platform creates more sensor processing challenges than most competitors.
- Other collision avoidance systems are more reactive, operating over much shorter timeframes (more like trajectory planners than path planners), whereas this system provides not just short term planners but longer range planning.
- The sensor fit on MAST is broadly similar to other products/projects.

Demo 5 (as noted in section 3.2 above) was the final deliverable of the project; a showcase of all the technological developments made to date. Each leg of the scenario demonstrated a particular element of the functionality that has been developed as part of the project.

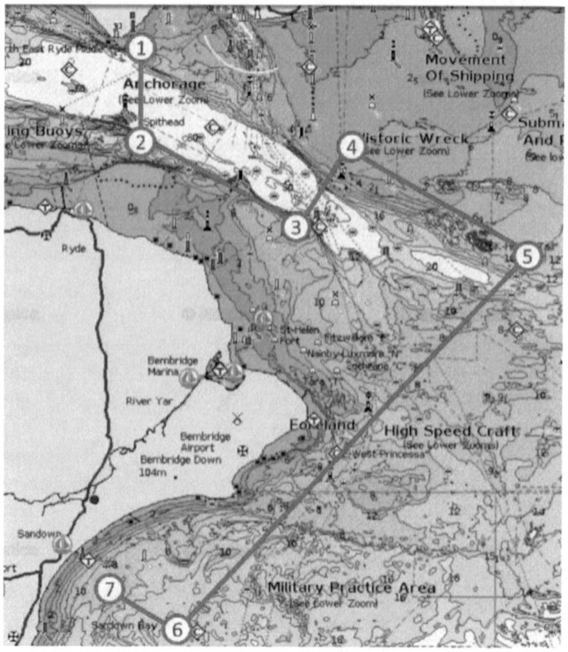

Figure 4: Overview of demo 5 operational area

The MAST USV was demonstrated operating autonomously (with full collision avoidance) from just outside the entrance to Portsmouth Harbour out to the outer approaches of the Solent, this included crossing the main shipping channel twice and operating at high speed. The Solent is one of the busiest shipping areas in the world and this was a significant achievement for the project.

Key functionality demonstrated during this run included various potential collision scenarios (including emergency evasive maneuvering), the last response engine, route planning, automatic sensing of the sea state and corresponding adjustment of the USVs speed. The demonstration also included a demonstration of the work done to date using machine learning techniques to automatically classify different types of vessels and buoyage.

4. CONCLUSIONS

MAST and the collision avoidance system is now at the advanced prototype stage. It is clear that the system now has a level of performance and assurance that provides the opportunity for two crucial exploitation testbed activities:

1. The modular architecture approach allows additional or alternative subsystems to be easily compared against the current baseline, to evaluate performance improvement. Given that it is now established it is fundamentally possible, research can be focused towards pushing the limits (e.g. operating closer in shore or in more challenging environments).

2. Integrating into the wider context. There are numerous higher layer mission management and situational awareness systems at various TRL. MAST now provides the opportunity for those systems to experiment with real-world uncertainty by integrating with an actual representative platform.

Dstl and ASV therefore welcome interest from other organisations who would like to explore and experiment with running their systems either on board MAST or otherwise controlling or commanding it.

5. AUTHORS BIOGRAPHY

Dan Hook CEng, MRINA is a qualified Naval Architect and Chartered Engineer with over 15 years' experience in the unmanned marine industry. Upon graduating from the University of Southampton, Dan joined Seaspeed Marine Consulting as a Naval Architect in 2002 before moving onto the position of Technical Director in 2007. During his time at Seaspeed, Dan specialised in unmanned platform and high speed craft design and consultancy.

Dan currently holds the position of Senior Director – Business Development at ASV Global, a company at the forefront of developments in unmanned marine systems for the Military and Defence, Offshore Energy, and Science and Survey Industries. Dan has built a vibrant team of naval architects, engineers and robotics experts, seeing the company grow from 3 people to over 80 in just six years. During this time the team have delivered over 80 Unmanned Systems to commercial and defence customers in the UK and all over the world.

As well as his role at ASV Global, Dan maintains an active role in industry. He is currently the Chairman of the Maritime Autonomous Systems (MAS) Council run by the Society of Maritime Industries, and is a member of the Solent Marine and Maritime Steering Group.

DATA MANAGEMENT FOR SMART SHIP OR HOW TO REDUCE MACHINE LEARNING COST IN IoS APPLICATIONS

Rodrigo Perez Fernandez, Universidad Politécnica de Madrid/Spain, rodrigo.perez.fernandez@upm.es
Arturo Benayas-Ayuso, Universidad Politécnica de Madrid/Spain, abenayas@alumnos.upm.es
Francisco Pérez-Arribas, Universidad Politécnica de Madrid/Spain, francisco.perez.arribas@upm.es

SUMMARY

Shipbuilding process, generates a lot of information and data, which a priori makes it seem impossible to have all this data in real time, but the new processors, simpler and smaller, with a good connection to the Internet, make it possible.
The data management is, however, only one side of the coin of the Internet of Ships (IoS). Energy efficiency is a fundamental aspect also in new devices that connect to the network.
But IoS not only covers the stages of design or production of the boat. Once the sensors are in the components whose information wants to monitor, we will be able to obtain information throughout the life of the ship.
IoS is presented as a solution capable of detecting when a component on a boat is close to fail and must be replace, when we take the boat to repair when we have to paint again, when corrosion has reached a certain limit ... and all this from our pocket tool and early enough to avoid late or unforeseen performances.
IoS reaches this sector to ensure profitable production, or safe, efficient and sustainable process for all types of fishing vessels, tugboats, tankers, charges, ferries, dredgers and oceanographic ...
Data management in Smart Ships, including collect, process, saving and third party distribution should be regulated, controlled and done in the most efficient and secure way for the ship owner. This can be accomplished based on a deep study of each part which is involved in close-to-real response systems and its machine-learning control unit. In these Systems, each parameter change generates stress and material fatigue, reducing its lifecycle.

NOMENCLATURE

CAD	Computer Aided Design
DDoS	Distributed Denial of Service
GPS	Global Positioning System
IoS	Internet of Ships
IoSS	IoS System
IoT	Internet of Things
IT	Information Technologies
OSC	Overall System Control
PLM	Product Lifecycle Management
RFID	Radio Frequency IDentification
RPM	Revolutions Per Minute
SSIC	Ship System IoT Cell

1. INTRODUCTION

If the naval industry wants to adapt itself to the new technologies it must to start with the entire design process of the product, in such a way that the product to be prepared for the new technology. The shipbuilding product has two well-defined and separated periods: the design phase and the operation life. These two elements can be sustained through working methodologies. Of particular note is systems engineering. Systems engineering appears to solve the problems that need to be addressed in sectors with high added value, such as products for defense or aerospace engineering, which are high-risk sectors that require a very good design and construction processes from very early stages to reduce costs and ensure planning and schedule within a budget. Systems engineering requires the use of more specific technologies in order to be applied effectively. In systems engineering, there are innumerable applications that can be used to provide solutions in the different stages of the life cycle of a ship.

Computer Aided Design (CAD) systems are traditionally used for the design and construction stage and in more recent times, the *Product Lifecycle Management* (PLM) solutions are emerging to provide valuable tools for product management. However, these tools require an enormous amount of information that must be provided during the ship design and construction project stages. Therefore, it is necessary to find a way to join these two technologies: CAD and PLM so that all the value mutually contributed in a project not to be lost.

Nevertheless, this is only a side of the problem. There are different technological trends appearing as the engines of the great changes that are affecting us. Among them, the *Internet of Things* (IoT) stands out because of the impact it has directly on the world we see and touch. How it affects or can affect this trend to the world of shipbuilding, and what can the software industry bring to enter the IoT to the shipbuilding industry.

First, we need to analyze what is actually the IoT nowadays and what are its challenges.

2. CHALLENGES OF THE IoT

We can recognize several areas where IoT presents problems that need a common solution: security, standardization, and scale challenge and business orientation.

We are still very aware of the recent *Distributed Denial of Service* (DDoS) attack by thousands of malicious software infected devices connected to the IoT. What could happen with a ship awaiting responses from internet in the middle of a storm with a load of millions of dollars on board or with thousands of people inside?. Security is the most important problem to be solved in a connected IoT ship. There are several issues related with this question: identification of devices, protection against attack, control of updates, redundancy, etc. There are still no guaranteed solutions to this issue.

The rapid growth of IoT has led to an uncontrolled growth of devices connected to the Internet through particular solutions of the manufacturers. Each device manufacturer has built their own IoT solution by connecting their hardware to their cloud server to service their requests. According to Ahmed Banafa wrote in (Banafa, 2016) [2], *"As the industry evolves, the need for a standard model to perform common IoT backend tasks is becoming more relevant"*. Moreover, it becomes necessary to have a greater standardization if it is desired that this connectivity is a pole of growth of current society.

Another issue can be known as scale challenge. The large variety of devices that have relationships with each other, but that have different workflows and of course different update times. If updating them freezes the operation of the system, the working time of the system would be significantly reduced. It is necessary to control the update cycles or perhaps to directly prevent them until it is decided that it is time to do it globally or selectively.

The success of IoT will not go hand in hand with the early adopters, but with the companies and this will only be possible if the initiatives have a clear component oriented to the business, that is to say to add a value. It is necessary to identify which of the initiatives provide a clear value to the business, but taking into account that this value can be in various terms and not all directly economic. The clearest ones have to do with the optimization of energy costs, fuel consumption, choice of routes, safety at sea, but also in the work itself inside the ship (Muñoz and Perez, 2017) [3].

One of the conclusions of all the analysis about the consequences of the IoT is that objects must be re-thought from the early design.

The fact is that ships are subject to harsh environmental conditions makes that any technological advance to be applied inside a ship has to take into account from the very beginning. Currently all the designs in the modern society are made by information technology tools, computer aided designs, supported with adequate databases and with lifecycle management of the information. To address the challenge of the IoT for ships, it is necessary a new concept, the *Internet of Ships* (IoS).

3. A NEW CONCEPT. THE INTERNET OF SHIPS

It is estimated that in 2020, 25 billion of devices will be connected to Internet (Kirsch, 2015) [5]. This revolution that began a few years ago has aroused enormous interest in all industries and in some of them already works with apparent normality.

Nowadays it is possible to order directly from our refrigerator as soon as it detects that we need our regular products or smart lamps that light up alone when needed lighting. The world go on steadily toward what will be undoubtedly one of the most important revolutions in the history of humanity.

We could define the IoT as consolidation through the network of networks a "network" that staying a multitude of objects or devices, that means, to connect all things of this world to a network, we are talking about vehicles, appliances, mechanical devices, or simply objects such as shoes, furniture, luggage, measuring devices, biosensors, or anything that we can imagine.

At its core, IoT is simple: it's about connecting devices over the internet, letting them talk to us, applications, and each other. But IoT is more than smart homes and connected appliances, however. It scales up to include smart cities *think of connected traffic signals that monitor utility use, or smart bins that signal when they need to be emptied* and industry, with connected sensors for everything from tracking parts to monitoring crops.

In this context the question is if the naval sector is ready for this revolution. Is it possible that this traditional and conservative sector moves into this technology? There is already evidence that the shipbuilding industry is no stranger to these developments and is already connected to the Internet some components of ships, as it is shown on [Figure 1].

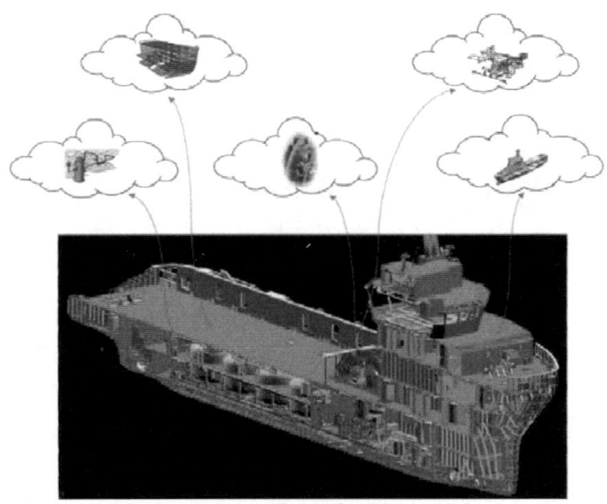

Figure 1. Ideal representation of a 3D model with access to the different ship design disciplines (Liu et al., 2016) [6]

As there smart home or smartphone, there are new smart ships that will be equipped with a network of sensors that capture a range of voyage information, including:

- Location.
- Weather.
- Ocean current.
- Status of on-board equipment.
- Status of cargo.

Ship owners can monitor the vessel's status in real time and apply analytics to current and historical data to make decisions that enable them to run more efficiently, saving time and fuel.

Sensors and *Information Technologies* (IT) are facilitating the introduction of new applications at sea, like energy distribution, water control and treatment, equipment monitoring in real time... The aim is to take this technological revolution also acting in the design and production phases in order to build efficient, safe and sustainable vessels.

In a decentralized sector, like naval, where often the engineering and production are in different locations and where critical decisions cannot wait, the IoS or connection through the network of critical components in the design / shipbuilding starts to glimpse as something that the sector cannot obviate.

The idea is to monitor all those parts in which early detection of events allows us to make the right decisions. In this sense, the available sensors during the early stages of construction of the ship, allow us to identify if the construction of the boat is completely according to the design we have created with CAD. If we can reduce materials or use another material, if we must change anything according with naval architecture calculations... The continuous monitoring integrated with a naval design CAD as FORAN will reduce costs and avoid mistakes and make decisions in real time from the shipyard, design offices or from remote locations.

Nowadays solutions CAD like FORAN can be used in a pocket tools, making it the indispensable ally in this new technological revolution, see [Figure 2].

Figure 2. 3D model on a virtual portable solution like tablets or smartphones (Alonso et al., 2017) [1]

Shipbuilding process, generates a lot of information and data, which a priori makes it seem impossible to have all this data in real time, but the new processors, simpler and smaller, with a good connection to the Internet, make it possible.

The data management is, however, only one side of the coin of the IoS. Energy efficiency is a fundamental aspect also in new devices that connect to the network.

But IoS not only covers the stages of design or production of the boat. Once the sensors are in the components whose information want to monitor, we will be able to obtain information throughout the life of the ship.

IoS is presented as a solution capable of detecting when a component on a boat is close to fail and must be replace, when we take the boat to repair when we have to paint again, when corrosion has reached a certain limit ... and all this from our pocket tool and early enough to avoid late or unforeseen performances.

IoS reaches this sector to ensure profitable production, or safe, efficient and sustainable process for all types of fishing vessels, tugboats, tankers, charges, ferries, dredgers and oceanographic ...

4. INTRODUCTION TO MACHINE LEARNING APPLIED TO SHIP. CHALLENGES AND LIMITS.

Machine learning is mainly associated to neural networks and deep learning. According to M. Nielsen wrote in (Nielsen, 2017) [4]:

- *Neural networks, a beautiful biologically-inspired programming paradigm which enables a computer to learn from observational data.*
- *Deep learning, a powerful set of techniques for learning in neural networks.*

These methods are commonly applied to image, speech or handwriting recognition methods. These methods can be applied for a full learning of a subject or to improve automatic correction in certain parts of a ship system. These corrections are made like operators but in real time and compensated inside the full system.

Neural networks are based on neurons or perceptron, as it is shown in the [Figure 3]:

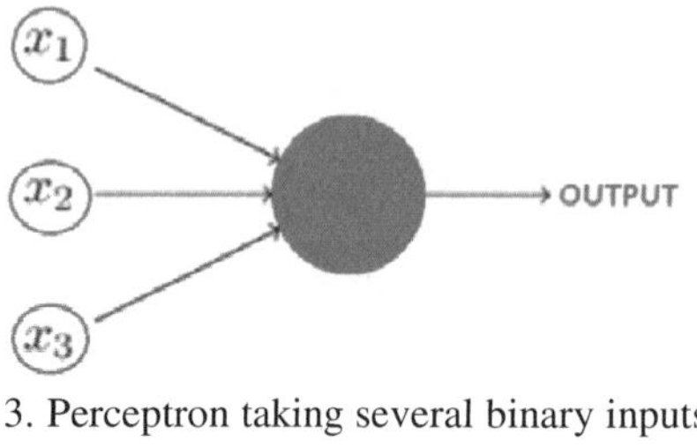

Figure 3. Perceptron taking several binary inputs and one binary output

These items several inputs and one or more outputs, depending on the expected result. One example of this is the exercise develops by M. Nielsen in it book about number recognition (Nielsen, 2017) [4] [Figure 4]:

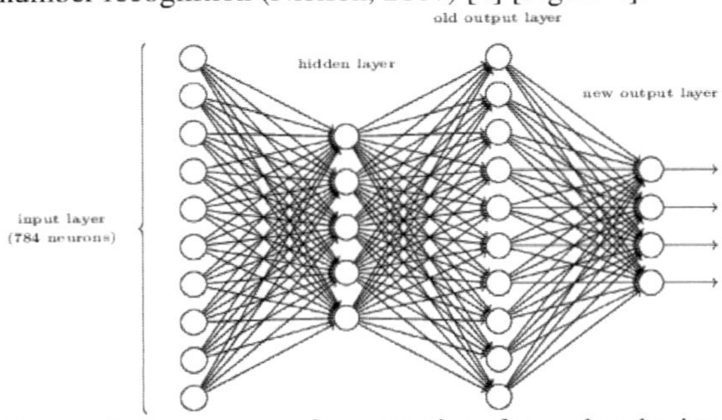

Figure 4. Neurons and connection for a handwritten number recognition

Each connection between neuron represents how previous neuron value affects to next layer neuron (weight), creating an influence pattern. Also, to correct this pattern of the full effect of all neuron connected to one, it is use a scalar number called bias.

The combination of weights and bias defines how our neural network works. Each neural network has an apprentice time, based on this learning we set and correct the network parameters, which are these weights and bias.

To neutralize influences of one layer to another, we can use a function to set the values of the next neuron between certain values, like the sigmoid function which establish the values between *0* and *1* for any result in the other layer of the network:

$$\sigma(z) \equiv \frac{1}{1+e^{-z}} \quad (1)$$

This function is useful for slow learning networks, because offers a smooth transformation of the values in each side. Most changing values are between *-2* and *+2*, representing *0* value of sigmoid the *0.5* of the activation level in each neuron. *z* function represents the activation propagation between un layer of neurons and a certain neuron connected in next layer, using weights and bias as corrections:

$$z = \sum_j w_j \cdot a_j - b \quad (2)$$

Where:

- w_j: weight of neuron j to affect next layer neuron.
- a_j: activation of neuron j to affect next layer neuron.
- b: bias correction of the result activation of next layer neuron.

For deeper information, refer to (Nielsen, 2017) [4].

All this parameters represents possible point of improvement of the neural networks, learning is based on corrections made in the weight and bias, also in the intermediate function, applying a different one from sigmoid, which can improve this learning, to obtain the expected result for correct inputs and a null result for incorrect inputs.

When an algorithm is set, cost function of a neural network, as the sum of the differences between expected result and obtained in last layer neuron normalized to one, represents how efficient is our algorithm against a test, and the weight and bias is our correction set of parameters.

Based on this, each group of elements of a ship system can be considered as an isolated cell, all the sensors and internal control unit can be programmed to react in real-time to internal stimulus, starting and element, stopping another, adjusting, etc., and finally communicating all the minor changes to a central unit in the ship.

Per each stimulus caused internally in the *Ship-System-IoT-Cell* (SSIC), normalized to *1*, the inner control takes a decision:

- Stable system part: do nothing.
- Adjust internally: correct one parameter which affects to the result of the neural network.
- Request an external adjust of any element out of the SSIC which directly or indirectly affects to this cell.
- Emergency: start secondary-emergency cell to avoid system failure, and stops securely all is internal components to minimize damage. This is direct connection, avoiding to wait for an external decision, even if this is taken by an operator or a control unit.

Any system of the ship can be divided in autonomous groups which are connected by a common element.

To avoid a fully abstract explanation, this article is going to be centered, as an example, in one of the systems of the ship, main engine refrigerating system.

This system is one of the principal help for the propulsion of the ship, even if is based in a purely fuel, diesel-electrical or nuclear propulsion, refrigeration is one of the main problems.

5. DIVISION OF SHIP SYSTEMS IN INDIVIDUAL IoT CELLS

Each cell needs to be autonomous, which is not the same as isolated, because in a ship all the items are related in a direct or indirect manner.

Steps to separate systems in SSIC:

- Divide each system in the smallest items-related groups. Each group can be considered as a module.

- Each group can be considered as a cell to be feed, and which is related with other system cells or even with other cells in other systems.
- Items should be enough close to have a direct relation.
- Adjusting parameters need to be also related. Any change in one affects to others.
- Add all elements an operator checks to adjust this group.
- Admit external signals as another control mechanism. This signal is always considered as inputs.
- Emit some external signals as inputs for other SSIC.
- Communicate status to main control system.

Developing these steps in previous example of sea water refrigerating system, in one of its cells. One of the groups which can be considered as a SSIC in the main engine refrigerating systems, usually sea water, is main pump group. This group usually for huge engines can have a configuration of *3* pumps, working in interchangeable cycles of *2* working and *1* remaining until required.

This SSIC is feed in two manners: sea water, as principal element, and electricity: high power to run the pumps & actuators, and low power to run the sensors and interchange information.

Main difference with common installation of this group or SSIC is the fact of it needs to be autonomous.

External input signals are:

- Start/stop command which can be translated as required flow in main engine for refrigeration. Input signal.
- Inlet or suction flow.

Internal input signals are:

- Remember which is the last pump started, controlling if this pump has complete a working-cycle or not, decide if it needs to started again.
- If last pump is restarted due to not a complete cycle is cover, begin to count the working time in an equivalent way as if it a non-stop group. This is one of the input parameter of our neural network.
- Check which valves are open and which are closed.
- Avoid "*water hammer*" in pumps and in solenoid valves. It is one of the most disrupting elements in pressure systems. This have a simple solution, the use of slow solenoid valves, but this behavior should also be included in our neural network.
- Each solenoid valve, due to previous points, represents two input parameters: first is open/close state and second time required to change state, because the slowness of the valves can be controlled.
- Non-return valves should represent another input parameter, with the open/close state.
- Check if pumps are primed or need to be prepared. Each prime state of a pump represents also a parameter in our neural network. This state can be represented as a yes/no or as a level. Second option has better options for regulation of our neural network.
- Each pump should emit also current *Revolutions Per Minute* (RPM). This is an input/output parameter of our neural network.
- Initially, an electronic thermometer should be installed in inlet and outlet to control fluid temperature, these two parameters are also inputs.
- Electronic pressure gauge, flow meter and thermometer should also be installed in the inlet and outlet of the cell. Input parameters.
- Electronic pressure gauge and flow meter should be installed in each pump outlet, after the output valves to control if pumps are in the expected pressure and flow required by external start order. Input parameters.

Internal outputs:

- Which pump is started and which is stopped.
- Which pump is in prime process.
- Which pump should increase its RPM.
- Which valves are in an opening process.

External output parameters, out of neural network:

- Inlet and outlet circuit parameters, pressure, temperature and flow.
- Pumps working time.
- Cell error codes, to be calculated in the Main Cell Controller which can be a PLC or any industrial programmable device.
- Start emergency cell to avoid full system failure. This is the unique neural network output.

Our network for this example has: 35 input neurons and 16 output neurons.

The hidden layers or intermediate layers are going to define the interactions between inputs and outputs, also, in this case, outputs affects inputs. Each output sets a change in the cell.

Main advantages of this simplification is that training this neural network cell can be do in laboratory, creating a set of working circumstances, from old values or any other situation, even with operators experience, improving the circuit and minimizing onboard impact.

Also, a set of errors can be reproduced and simulated on workshops during the training improving the parameters of this neural network and reducing human impact in the usual work of the cell.

More inputs and outputs can be add, improving the quality of this cell and the set of detected, autonomous corrected and avoided errors.

Last but not less important, electronic neural network represents also a piece of the system which can cause the error, due to this, it is important to set a group of zero test which helps ship machine engineers to identify if problem should be localized in any sensor or in the neural network itself develop by deep corrections.

If neural network in some circumstances makes the SSIC fails, with this test data can be restored to its default values, and all data collected till failure can be transmitted also to the shipyard in order to study and correct the issue.

6. SHIP SYSTEMS LIKE SHIP IoT ORGANS

By dividing each systems in autonomous, auto regulated and interrelated IoT cells or SSIC, reduces system overall failures and prevent problems in other parts of the ship which are not directly related to system under in issue circumstances.

Refrigerating systems can be divided in SSIC cells as follows, for example:

- Sea water suction.
- Main pump group.
- Secondary/auxiliary/emergency pump group.
- Main engine circuit.
- Sea waters discharge.
- *Overall System Control* (OSC).

Each SSIC generates inputs for other parts of the system, and globally, can collect from all parts of the system in the engine control room, where the OSC shows in a easy way the status of each cell and, also, some data can be distributed to the main bridge control console.

In this way, each OSC represents an "organ" of the ship as a summary of all the cells of the system represented by its shared information.

Each organ have a relation with others by OSC communications. Following with our example, refrigerating system full failure can communicate with fueling and lubricant system, stopping main engine to avoid a bigger problem. Also with main bridge to give an alarm, send a report to the ship owner and communicate with the nearest port ship last *Global Positioning System* (GPS) location.

With all this information, engine room engineers can start reparation process, or wait if problem cannot be solved with onboard tools and materials.

Extending the biological concept of "system" to the ship, all system related direct or indirectly with the propulsion can be an *IoS System* (IoSS) which groups all these organs to produce a simple action in the ship.

Creating bigger groups helps to classify errors, problems and also improves maintenance, due to each simplification requires a shorter list of spare parts.

Last organ is represented by the "brain" of the ship, this is the main OSC which recovers information from all IoSS, their organs and SSICs.

7. CONCLUSIONS.

As an example of IoS, the connectivity in smart ships will be extended to the commercial mission to act autonomously in operation conditions. A commercial vessel can transmit its navigation situation or its loading conditions.

All these means a huge amount of information to be managed and analyzed. New programs have to be developed to obtain the best use of such information so that the design can be improved from real function information of the design and it can be self-maintained with the connection with this huge cloud information to create method that the objects can achieve certain intelligence.

The growth of the IoS is linked to the increase of information and the management of Big Data, with the property that somehow IoS identifies information and direction and order to a specific purpose, while the concept of Big Data is more generic.

Treating ship like an alive being, and trying to simulate its parts such as organs or cells helps to introduce a neural network learning which can adjust each SSIC to work in the best conditions extending the life of its parts, preventing from damage other parts of the ship and helping to make a preventive maintenance of each part.

Neural networks represents an automated adjust of each simple group of things, because its parameters can be set to absorb external interferences.

Other advantage is that this systems can be trained individually, creating common, rare and unusual conditions.

Last advantage is, if something happens in a sailing ship, it can be reproduced in the shipyard based on this model, and getting a full condition simulation. This helps to focus the problem.

Adding some *Radio Frequency IDentification* (RFID) tags to the parts, can improve system information in ship brain, and if some parameters can be improved due to new known experiences can be downloaded in the ship to improve SSIC behavior.

8. ACKNOWLEDGEMENTS

The authors would like to express sincere gratitude to SENER for providing the support and information for this paper.

9. REFERENCES

[1] V. Alonso; R. Pérez and A. Valderrama, «World-class electrical design tool for shipbuilding» SMC 2017. Houston, TX, USA, October 2017.
[2] A. Banafa, «IoT Standardization and Implementation Challenges» Newsletter IEEE Internet of Things, July 2016.
[3] J.A. Muñoz and R. Pérez, «The Internet of Ships: a new design for Smart Ship» The Naval Architect magazine, January 2017.
[4] M. Nielsen, «Neural Networks and Deep Learning» On-line free book, December 2017.
[5] D. Kirsch, «The Value of Bringing Analytics to the Edge» Hurwitz & Associates. 2015.
[6] G. Liu; R. Pérez; J.A. Muñoz and F. Regueira, «Internet of Ships: The Future Ahead» World Journal of Engineering and Technology. Vol. 4. N. 3D. 2016.

10. AUTHORS BIOGRAPHY

Rodrigo Perez Fernandez is the head of Naval Shipbuilding Business Development in SENER. He is professor in several subjects at the Marine Engineering School of the Technical University of Madrid.

Arturo Benayas-Ayuso holds the current position of CAD & PLM Senior consultant at SENER Ingeniería y Sistemas S.A. He is PhD student at the Marine Engineering School of the Technical University of Madrid.

Francisco Perez-Arribas he is also professor in several subjects at the Marine Engineering School of the Technical University of Madrid.

NO TIME FOR DOWNTIME

I Kelsall, 3M United Kingdom Plc, UK

SUMMARY

With the emergence of big-data and automated shipping systems within the maritime industry the potential for disruption to service as a result of even a small fire needs to be considered by operators. Even more than in the data centre industry where disruption of service represents a loss of data or productivity, fire on board automated shipping could represent a loss of control of the vessel itself.
Selection of fire protection systems should be designed to minimise the potential for loss of data and loss of productivity. Comparison to the datacentre industry can be made and fire protection systems can be designed to reduce the potential for downtime to a minimum.

1. INTRODUCTION

A comparison may be drawn between today's data economy and the growing expansion of smart technologies across all industrial sectors. The digitisation of the shipping industry means that new risks emerge depending on the reliance on technology. Comparing the major causes of failure across the different industries reveals common trends and highlights that similar mitigation of risk may be effective.

2. MAJOR RISKS AFFECTING SHIPPING TODAY

Safety at sea is a clear priority for the maritime industry and is subject to continual monitoring to understand the biggest risks to seafarers, vessels and commercial operations. Safety is addressed by regulation such as the SOLAS convention, however regulation alone cannot address and prevent all incidents that occur at sea. Between 2011 and 2016 there were 16,539 marine casualties or incidents including 600 fatalities recorded by European member states [1].
Data collected by the European Maritime Safety Agency (EMSA) on Marine casualty events with a ship categorises events as follows:

- Missing
- Loss of Control
- Hull Failure
- Grounding/Stranding
- Flooding/Foundering
- Fire/Explosion
- Damage to ship or equipment
- Contact
- Collision
- Capsizing/Listing

Of all the casualty events, 50% were considered as navigational casualties including the categories of Contact, Collision and Grounding/Stranding. Furthermore, when looking at the contributing causes of all marine casualties over 70% were found to be attributable to human erroneous actions with failure of supervision the most common factor.
The next highest contributing factor group for marine casualties after human error is Equipment Failure with maintenance representing the major factor within that category accounting for over 20% of marine casualties.

Fire and Explosion on board vessels accounted for approximately 5% of incidents between 2011-2016. In comparison with other event types this is a relatively rare occurrence; however the potential consequences of fire at sea mean that this is still a significant threat that needs to be addressed.
Up to 50% of all Fires on board originate in the engine room with 70% of all engine room fires caused by oil leaks from pressurised systems. [2] The majority of all other fires start in the cargo hold or accommodation. With a significant proportion of fires affecting the ships propulsion and control, effective management of fire hazards is essential to ensure that the risk is minimised.

3. CHANGING RISK HORIZONS DUE TO EMERGENCE OF NEW SMART SHIP TECHNOLOGIES

Smart ship technologies can bring potential benefits for safety at sea in a number of ways. Fire safety management needs to consider the changes that arise due to the emergence of these new technologies and be adapted to be fit for purpose in the future.

3.1 AUTONOMOUS SYSTEMS

Remote controlled and subsequently autonomous systems will see significant benefits through the removal of human error from the decision-making process.
Navigational casualties caused by failure of supervision or fatigue will be significantly reduced by introduction of new technology and will also have potential to access more remote or dangerous shipping routes due to the reduction of risk to crew. Whilst autonomous shipping systems will reduce the potential for human error at an individual ship level, interactions between the unmanned, autonomous vessels and manned vessels will still need to be addressed.
Removal of trained sea-farers from all decisions may not be entirely positive. Human intervention in emergency or unexpected situations or challenging routes may still be expected to have a positive role to play as seen in the Airline industry where pilots are still required for take-offs, landings and emergencies despite an almost entirely automated process.

Human intervention currently has a major role to play during the response to any fire with firefighting and fixed extinguishing systems requiring manual activation in most cases. Although automatic fixed fire-fighting systems are available commercially, regulation changes would be required to allow these to be used in many cases and the solution would need to address the potential for human access to protected areas at some point.

3.2 SENSORS AND PREDICTIVE MAINTENANCE

Another smart ship development with the potential for improving safety and reducing the risk of fire hazards on board is the introduction of sensor technologies and networks for managing the ships condition and implementing a predictive maintenance and operation system.

With Equipment failure being the second largest cause of marine casualties within the European Union and failures in maintenance being the leading factor within this category it is clear that safety would be an area that benefits from the development of smart ship technologies. Predictive maintenance systems and sensors for measurement and analysis of performance will improve the reliability and efficiency of systems on a daily basis but also reduce the potential for critical failures. Furthermore, the ability of equipment manufacturers to gather aggregated 'big data' from a range of industries and operating conditions will allow weaknesses and defects to be engineered out of systems further improving reliability in the long run.

With Oil leaks on to hot surfaces being the leading cause of engine room fires developing an understanding of the risk scenario in relation to overall system may be an avenue to explore. However, application of high cost predictive maintenance to low cost, easily replaced equipment may be limited. Oil leaks may originate from a variety of issues such as:

- Loose joints
- Fractured pipes and mechanically damaged (perforated) pipes on both high and low pressure fuel lines
- Bleed cocks on generator fuel filters working loose
- Pipe unions that are over or under tightened
- The fracture of flange bolts if over tightened
- The fracture of cyclically stressed bolts or studs those are under-tightened, such as those securing fuel injector pumps
- The use of unsuitable seals or gaskets which deteriorate due to the effects of heat
- The rupture of high pressure oil and hydraulic fluid hoses due to mechanical damage or aging. [3]

Unless the overall system performance can be analysed to identify loss of function resulting from small leaks it would be unlikely that such causes would be directly addressed.

On the other hand, hot surfaces arising in industrial processes are already addressed through the use of predictive maintenance analysis techniques such as Infra-red thermography to identify faulty circuits, leaks from sealed vessels and other faults leading to the generation of heat. Elimination of a potential ignition source through sensing and analytics could be an effective fire safety management benefit.

3.3 ENERGY STORAGE AND MANAGEMENT

Another technology area likely to have an impact on fire strategy and risk management are advances in energy storage. Improvements in battery technology and chemistry have allowed significant increases in energy density and Energy storage systems based on Lithium ion cell technology have become a viable option for marine applications in both electric and hybrid propulsion systems. Applications such as passenger transport are particularly suited to these technologies because the regular need for docking allows recharging of batteries on a routine basis.

Energy Storage systems have many operational benefits. Key features offered by such systems include absorption of load variations such as peak shaving and dynamic performance enhancement or ensuring continuous operation of power through spinning reserves or uninterruptable power supplies (UPS). Of course, sustainable zero emission operation is also possible with such system depending on the power requirements and size of the vessel. It is possible to operate fully electric, short distance transportation or operate larger vessels in port or whilst docked to minimise emissions in coastal areas.

The main factors preventing the growth of lithium-ion battery includes capacity fade, and possible occurrence of thermal runaway [4].

Battery development continues at pace in order to address these performance and safety concerns. New electrolyte chemistries, elimination of liquid electrolytes through the use of polymer or solid-state electrolytes and supercapacitors all represent potential next generation solutions and will reduce the probability of any thermal runaway. But what is the probability of failure with Lithium ion cells?

Lithium Ion batteries are generally very reliable thanks to extensive regulation and technology such as battery management systems to limit the failure rate in the order of 1 in 10 million cells [5]. However, with global Li-ion cell production in the region of 6 billion units in 2015 [6] there remains potential for a considerable number of failures even without accounting for misuse and abuse.

For the growing use on board ships it is therefore worth accounting for potential thermal failures within fire management practice.

4. DRAWING A PARALLEL FROM EXISTING DIGITAL INDUSTRIES

As shipping becomes more reliant on data, whether through autonomous or remote control systems or and Internet of Things (IoT) approach to remote maintenance and inspection, parallels with the data-centre sector can be drawn. Each smart ship potential becomes an edge computing facility running a number of processes related to a number of core activities that need to function securely, remotely and with minimal latency that may be associated with cloud computing.

4.1 THE IMPACT OF FAILURES ON THE DATA INDUSTRY

Critical systems failures within the data industry will not typically have catastrophic results that could be experienced at sea. Loss of life or protection of the structure of the data centre can be managed through good practice and well-designed systems. However, the potential impact of data centre failures to business continuity or critical processes is huge.

The business cost of data centre outages has been estimated at $8,851 per minute with the average shutdown (partial or total) lasting 95 minutes [7]. The consequences of these outages for business are typically business disruption, loss of business or end-user productivity. In August 2016 a fire at a Delta Airlines Datacentre led to the cancellation of approximately 2300 flights and cost the company in the region of $100 million [8]. Whilst this example is not directly life threatening the impact is far reaching for businesses and their customers. If downtime impacts end-user productivity, which could represents managing a critical process such as hospital scheduling and diagnostics or navigation for smart shipping, then the potential outcomes could be more severe than loss of revenue.

4.2 KEY FAILURE MODES WITHIN THE DATACENTRE

It is possible to draw parallels between the failure modes that most commonly affect datacentres and those that result in marine casualties in the shipping industry today. Human error and equipment failures account for 83% of all unplanned outages with weather and 'other' factors accounting for the remainder of outages however human error represents a much lower individual portion.

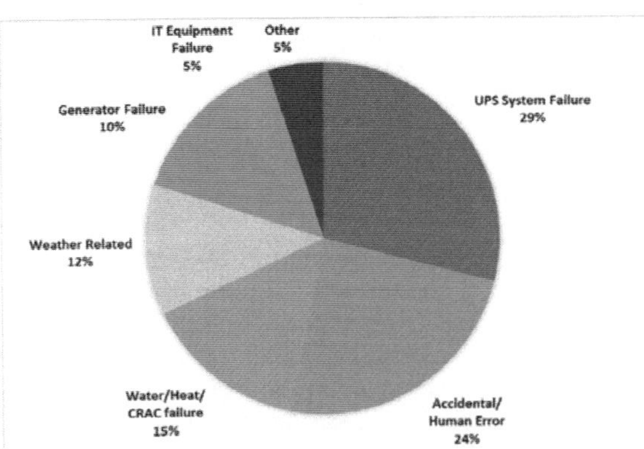

Figure 1. Primary Root causes of unplanned datacentre outages [9]

In this case fire is typically seen as a secondary event due to the failure of electrical or power equipment within a datacentre it is therefore difficult to define the likely occurrence of a fire and the sensitivity of declaring such high-profile incidents by datacentre providers or firms means that under-reporting is also possible.

4.3 MEASURES TO IMPROVE 'UPTIME' PERFORMANCE

With the data business relying on continuous availability in order to survive, measures to ensure that uptime targets are achieved are widely used and leveraged as a business credential. Various accreditation schemes also exist for operators and locations based on the ability and understanding of systems needed to ensure business continuity.

We will look at some of the main considerations when designing datacentres to achieve robust performance

4.3 (a) Prevent System Failure

As we have seen equipment failure is responsible for a large proportion of complete outages. Prevention relies largely on 2 key factors that are closely related to developments within Smart Ship design: Redundancy and Monitoring & Maintenance.

Achieving the correct level of redundancy is dependent upon the needs of any given organisation and the critical nature of the data or IT process that is occurring within the datacentre. Redundancy can be built into the entire process however complexity of the overall system increases as each level of redundancy is built in. With a business based on data the initial level of redundancy is to have a "failover" where the data process can be switched to another facility. Where the data is running a critical process such as the case with smart shipping the likelihood of this being achievable is reduced.

Redundancy is widely used for power supply and cooling/heating with datacentres designed to achieve the required level for the business. When designing the redundancy level factors such as required availability (24/7 operation for example) and system reliability will influence the design greatly. An unreliable system operating all of the time will have a high likelihood of failure and redundancy requirements would be excessive. Availability and Reliability also become important for the monitoring and maintenance of the system. In order to ensure or potentially increase reliability it is possible to design systems that are constantly monitored and design maintenance plans based on the output when related to the availability requirements. Once again this is very closely related to developments in Smart Ship Technologies.

4.3 (b) Reduce Human Error

Reduction of Human error is another area of concern for datacentres and a major component in the assessment and measurement of datacentre performance and

accreditation. At a basic level ensuring an adequate level of staffing and training in place at a datacentre can minimise the impact of a component failure by allowing an immediate response. However, this needs to be supported by adequate policies and procedures to ensure that over-reliance on individuals does not lead to failures in the absence of a given expert.
The ultimate goal for datacentres to reduce human error will also lead to automation and the use of artificial intelligence or robotics to perform both manual tasks and software update tasks.

4.3 (c) Robust Disaster Planning

Disaster planning has the same objectives as reducing human error and ensuring equipment reliability. It however takes into account the need to achieve these goals in response to an external threat (such as extreme weather events or accidents). Rather than being a stand-alone tactic it is about ensuring that systems will work and that people involved know what to do during an event. Carrying out drills for system start-ups and shut downs can test both the abilities of the equipment and personnel and should be part of any plan to ensure that smart ship technology can provide continuous performance.

5. CONSIDERATIONS FOR ESTABLISHING EFFECTIVE FIRE SAFETY STRATEGIES FOR SMART SHIPPING

Given the rise in smart ship technologies, it is worth considering how the approach to fire safety strategies and protection systems may be affected and how systems may allow for adaptations arising from new technology.
The fire safety objectives for ships are clearly outlined by SOLAS II-2 regulation [10] as follows:

- Prevent the occurrence of fire and explosion;
- Reduce the risk to life caused by fire;
- Reduce the risk of damage to the ship, its cargo and the environment;
- Contain, control and suppress fire in the compartment of origin; and
- Provide adequate and readily accessible means of escape for passengers and crew.

Each aspect of these objectives will vary to some degree depending on the vessels design and purpose. We may also see the rise of new Smart ship technologies affect the objectives in both positive and negative ways.

5.1 PREVENTING THE OCCURRENCE OF FIRE

Smart ship technologies will have the potential to both introduce new risks whilst potentially reducing some existing risks. As previously discussed intelligent ship systems capable of monitoring and sensing of critical engine room processes may hold the potential to reduce the number of oil leaks leading to engine room fires through improved maintenance planning.

On the other hand introduction of new technologies and increased computing will potentially create additional ignition sources and fire loading. Expanded computing and control spaces will need the application of regulation to ensure that the risk of fire is adequately mitigated and the use of new power sources such as Lithium Ion battery technology may need new requirements for protection systems to ensure adequate containment and prevention of thermal runaway.
Automation also creates the potential to reduce the occurrence of fire due to human error. Mishandling of flammable materials, introduction of ignition sources during hot work or illicit smoking, interference with firefighting systems are all potential risk scenarios created by the introduction of humans. Whilst this can be managed through training and procedures it cannot always be eliminated

5.2 REDUCING THE RISK TO LIFE, THE SHIP, THE CARGO AND THE ENVIRONMENT

With Autonomous Unmanned Ocean-going ships predicted to be a reality as soon as 2035 [11] there is a growing opportunity to enhance safety. It should be considered that automation may allow crew to be removed from tedious, unskilled positions that pose a safety risk to that person and reduces the overall number of people on board at risk from fire. The focus in that case shifts towards protecting the vessel, the cargo and the environment and defining fire protection measures appropriate to that purpose whilst also being capable of remote operation.
Reducing human error is one major safety benefit that has already been discussed however human interaction and decision making during emergency situations can also prevent incidents from occurring. Decisions on how and when to respond to different fire scenarios that have historically been taken by crew would need to be automated by systems that are capable of those same decisions. Whilst automated systems exist today they rely on correct design, installation and maintenance to operate at the appropriate times. Failure to operate soon enough could lead to extensive fire spread and damage, whilst overly sensitive system designs could operate too often leading to excessive cost and potential damage resulting from fire suppression and loss of protection following any event.

5.3 CONTAIN, CONTROL AND SUPPRESS THE FIRE IN COMPARTMENT OF ORIGIN

Fire protection systems design is well established for today's ships and in many ways, will continue to be adequate as smart ship technologies emerge. However, It is necessary to understand the major characteristics for fire protection options and how these properties could be leveraged depending on the technical developments that arise.

5.3 (a) Passive Protection

Passive Fire Protection is routinely used to contain fire to its original compartment and prevent spread of fire throughout the shup Maintaining the integrity of boundaries between subdivisions relies on insulation and protection of any penetrations that may be placed through these divisions.
Whilst largely unaffected by smart ship technologies there may be potential for some changes in the design of vessels. The potential to remove accommodation space to increase cargo space or increase efficiencies in design may move these boundaries but will not eliminate them.

5.3 (b) Detection and Alarm Systems

Detection and Alarm systems should be designed to raise awareness to the location of a fire at as early a stage as possible. Early identification of an incipient fire allows for rapid response for containment of fire and smoke and potentially extinguishment before the risk of damage to critical equipment becomes too great. In the case of manned vessels detection and alarm also allows for an appropriate response by crew or passengers in accordance with the fire safety strategy. With automated shipping, these systems will be responsible for the activation and monitoring of the firefighting response by the remote operation and control centre.

5.3 (c) Suppression Systems

Fixed Firefighting systems for machinery spaces and data/control rooms need to be designed around the objectives of the equipment that is being protected. Systems that are critical to the overall safety of a vessel (such as propulsion or navigation) need to be protected by a firefighting system that allows for continued operation and minimal damage resulting from the extinguishing agent in addition to the fire itself. Aerosols, foams and water based extinguishing systems all have the potential to leave potentially damaging residue on electronic equipment leading to extensive clean up or damage. Water mist systems are now extensively used for engine rooms. By comparison when looking at server/control rooms the recommendation for the construction of datacentres would be to only use water based systems for the protection of the building itself. When it is essential to reduce equipment damage from an incipient fire to minimum possible levels, or to facilitate the return to service, a clean agent fire extinguishing system with detection to protect the data processing equipment would be recommended [12]. Clean agent gas suppression systems are also well characterised for marine applications with a variety of agents available including Carbon Dioxide, Inert Gases and Halocarbons such as 3M™ Novec 1230.
Speed of detection and suppression is critical to minimise the potential for damage to data processing equipment and reduce the disruption to key processes. Fixed firefighting systems in the absence of crew should be capable of extinguishing incipient fires rather than suppressing fire as there is limited possibility for emergency response. Water based systems have historically been designed to activate in response to larger established fires and to reduce hot surface temperatures. They are also capable of controlling small obstructed fires but are limited in their ability to extinguish such fires [13]. For small obstructed fires arising in e.g. server cabinets, gas suppression systems would be considered more effective and are required by design to reach extinguishing concentration in 10 seconds (for Halocarbon systems) and 120 seconds (for Inert Gas) [14]. These speeds allow for fast extinguishment with minimum damage or downtime for equipment.
Human safety is a consideration for all fire suppression systems. For unmanned ships it is possible to consider the possibility of inserting the machinery and control equipment spaces with an inert gas such as Nitrogen or Carbon Dioxide and reducing the oxygen concentration level to below the point at which ignition cannot be achieved. This needs to be balanced with the needs for personal access to machinery spaces and the fitness of crew entering such spaces for an extended period. However the cost of equipment and energy needed to control and maintain such systems would also need to be considered.
Inert gas suppression systems used to flood protected space rely on achieving approximately a 12% oxygen concentration using gases that would allow safe access for limited periods of time. Carbon Dioxide however, will lead to unconsciousness and subsequently death if exposed to for more than 1 minute [15].
Halocarbon based fire suppression systems operate at lower design concentrations as their extinguishing mechanism is based upon cooling of a fire rather than reducing the oxygen concentration. This lower concentration means that asphyxiation is not a consideration. Cardiac sensitization is a possibility with Fluorinated compounds so comparison of the required design concentration with the assigned No Observable Adverse Effect Limit (NOAEL) for an agent is necessary to ensure that a design is safe.
Another safety concern is the familiarity of crew responsible for the operation and maintenance of fixed firefighting systems. In 2016 the highest recorded deficiency category for port state control was Fire safety [16]. Mishandling of such systems can lead to accidental release of extinguishing agents when machinery spaces are occupied. Whilst unmanned vessels will reduce the amount of time that protected spaces are occupied it may become more likely for routine maintenance to be carried out by crew that are unfamiliar with the implications of accidental activation of such systems.
Another consideration in the selection of fixed firefighting systems is the overall design of ships. Unmanned vessels have the potential to maximise efficiency by removing un-needed space such as accommodation space. Firefighting systems require differing amounts of storage space for extinguishing agent depending on the size of the protected space and

design concentration. Systems that require less agent give greater flexibility to designers to offer the same level of protection without requiring additional space or even separate storage rooms.

6. CONCLUSIONS

It is clear that extensive changes for the maritime sector are expected in the next 20 years as smart ship technologies emerge. Understanding the impact that these changes will have on the overall safety of shipping is of vital importance.

Technology in many ways will introduce many improvements for safety. With this, control systems and onboard digital technology becomes ever more critical for the overall control of vessels. Comparison can be drawn with the importance of datacentres in the growing data economy and the impact that downtime has on commercial operations. The potential consequences of downtime for smart shipping should be considered carefully.

Fire protection in Datacentres has become a standard part of the design with relevant systems selected based on the overall objective of the facility or data owner. In cases when there is 'no time for downtime' correct selection of a fire protection system is vital.

7. REFERENCES

[1] 'Annual Overview of Marine Casualties and Incidents 2017', European Maritime Safety Agency, 2017

[2] [3] 'Risk Focus: Engine room fires', UK P&I Club; Burgoynes Consulting Scientists and Engineers, 2016

[4] Ramadesigan, V. et al. 'Modelling and simulation of lithium-ion batteries from a systems engineering perspective'. Journal of The Electrochemical Society 159, R31–R45 (2012).

[5] Mitch Jacoby, 'Assessing the safety of Lithium ion batteries', Chemical and engineering News, Vol 91, Iss 6, 11th February, 2013

[6] Christophe Pillot, 'Rechargable battery market and main trends 2016-2025', Avicenne energy, 20th March 2017

[7] 'Cost of Data Centre Outages', Ponemon Institute, 2016

[8] Tanya Powley, 'British Airways counts cost of outage disruption', Financial Times, 28th May 2017

[9] Kevin Heslin, 'Primary causes of unplanned outages', Uptime institute Journal,

[10] 'Guidance to SOLAS Chapter II-2', Marine and Coastguard Agency,

[11] 'Safety and Shipping review', Allianz Global Corporate and Security, 2017

[12] 'Property Loss Prevention Data Sheet 5-32 for Data Centers and Related Facilities', FM Global, April 2017

[13] National Research Council. 'Fire Suppression Substitutes and Alternatives to Halon for U.S. Navy Applications'. Washington, DC: The National Academies Press. https://doi.org/10.17226/5744. 1997.

[14] 'Revised Guidelines for the approval of equivalent Fixed Gas Fire-Extinguishing systems, as referred to in SOLAS 75, for machinery spaces and cargo pump-rooms', IMO, MSC/Circ.848

[15] 'CO2 Physiological Hazards "Not just an asphyxiant!".' European Industrial Gas Association, Safety Information 24/11/E

[16] 'Port State Control Annual Report', Paris MoU, 2016

8. AUTHORS BIOGRAPHY

Ian Kelsall, holds the position of Application Development Specialist for 3M United Kingdom Plc as part of the Electronic Materials Solutions Division. He is responsible for the technical leadership and development of new applications related to Active Fire Suppression solutions from 3M. He has spent 11 years working on Occupational Health and Safety and Fire Safety across research and development and technical service. He is currently an Affiliate Member of the Institute of Fire Engineers.

Riding the wave of IoT: Strengthening regulations, improving safety and implementing Safeship®

C S P Hunter, A Saw, C Kay, and C M E Hunter, Coltraco Ultrasonics, UK

SUMMARY

International Maritime Organisation (IMO) Safety Of Life At Sea (SOLAS) is the critical code the industry at large wishes to see implemented. Beyond saving human lives, this protects the cargo and vessel. The IMO regulations developed further to protect the wider marine environment and today aim to make efficiencies. We understand that the "Smart Ship" can be widely interpreted from developing automated vessels to unmanned vessels, through to what this paper argues which is delivering a safe ship using innovative smart technology to increase safety at sea. Using ultrasonic technology, the Safeship® is achieved by the continuous monitoring of safety systems that are designed to combat the two main causes for vessel loss: sinking and fire. Using our expertise in ultrasonic technology, this paper argues that the Safeship® can be achieved through the monitoring of fixed gaseous extinguishing system contents and watertight integrity, benefitting from the continuing and developing technologies we can all ride the wave of the Internet of Things (IoT).

1. INTRODUCTION: The Smart Ship is a Safe Ship

Through ongoing technology advancements, ship owners can protect their crew, vessels and cargo with smart Safeship® technologies. This paper argues we can better achieve the Smart Ship, implement existing regulations, and strengthen new ones, by introducing more sensor monitoring technologies which will be linked by IoT. The application of the IoT and ultrasound allows the Smart Ship to be achieved here and now whilst research and development presents future opportunities. Continuous monitoring solutions benefit the crew: by being easy to operate, quick, accurate and better than traditional monitoring techniques, means the likelihood of their use and thus protection is much higher. Whilst regulations can sometimes be said to drive up costs in the short term, their longer-term impact is designed to save time, costs and improve the industry. Owners, managers and operators in particular must strive towards the Safeship®: holistic protection of their fleet.

2. FIRE

2.1 WHAT ARE GASEOUS EXTINGUISHING SYSTEMS AND WHY DO THEY LEAK?

To address the issue of fire in ships, which is one of the top reasons for vessel loss, the authors of this paper applied their expertise to an important problem in gaseous extinguishing systems: they leak. Gaseous extinguishing systems protect important infrastructure against special hazards, fundamental for the safeguarding of critical facilities. A ship's gaseous extinguishing system typically comprises between 200 and 600 cylinders each containing 45KG of CO_2 under high 720 psi/ 49 bar pressure. (Other suppressant clean agents such as FM-200® and Novec™1230 are becoming more widely used.) One of the highest probabilities of discharge occurs during their maintenance. Some marine service companies estimate that 20% of a ship's CO_2 cylinders have discharged or partially leaked their contents at some point in their lifetime. Taking CO_2 systems through as an example, although random checks may be suitable in some sectors, it is worth remembering that because the normal design concentration of CO_2 of 34-72 v/v % is above the nearly immediate acute lethality level, these systems have an extremely narrow safety margin. As these systems work through oxygen dilution rather than the chemical disruption of the catalytic combustion chain (which is the case with other clean agents), insufficient oxygen levels during an accidental discharge may allow a situation to spiral out of hand. Yet although this poses high levels of risk to the service companies and the crew, because gaseous extinguishing systems are highly pressurised, the risk of leaking and discharging is accepted as part of their use and this is shown in the regulations that demand their upkeep.

2.2 WHAT ARE THE REGULATIONS THAT GOVERN GASEOUS EXTINGUISHING SYSTEMS?

The implementation of regulations through Safeship® technologies ensure that the gaseous extinguishing systems are functional in the event of a fire. The regulations that currently govern gaseous extinguishing systems are the IMO SOLAS Fire Safety Systems (FSS) Code and the BS EN ISO 14520 standards.

IMO SOLAS & FSS Code Chapter 2.1.1.3 - "Means shall be provided for the crew to safely check the quantity of the fire extinguishing medium in the containers."

Clean agent fire-fighting systems are guided by BS EN ISO 14520 Gaseous Fire Extinguishing Systems. BS EN ISO 14520 -1:2015(E) assumes that the execution of its standards is entrusted to people qualified and experienced in the specification, design, installation, commissioning, testing, approval, inspection, operation and maintenance of systems and equipment, and who can be expected to exercise a duty of care to avoid unnecessary release of extinguishant.

ISO 14520-1:2015(E) specifically states in 6.2.4.2 Contents Indication that - Means shall be provided to indicate that each container is correctly charged and in 9.2.1.1 At least annually, or more frequently as required by the authority, all systems shall be thoroughly inspected and tested for proper operation by competent personnel.

In ISO 14520-1:2015(E) 9.2.1.3 - The storage container contents shall be checked at least every six months as follows: a) Liquefied gases: for halocarbon agents, if a container shows a loss of agent in quantity of more than 5 % or a loss of pressure (adjusted for temperature) of more than 10 %, it shall be refilled or replaced. b) Non-liquefied gases: for inert gas agents, pressure is an indication of agent quantity. If a container shows a loss of agent quantity or a loss of pressure (adjusted for temperature) of more than 5 %, it shall be refilled or replaced.

2.3 WHAT DOES THIS MEAN FOR THE CREW?

The crew have a responsibility to implement the regulations via regular testing, which enabled through smart ultrasonic technology and IoT, should be done continuously to avoid negligence and unnecessary risk. Misunderstanding exists across parts of shipping industry regarding the application of a part of the IMO SOLAS FSS Code: the need for crew to test the contents of their CO2, FM-200® & NOVEC™ 1230 Gaseous Extinguishing Systems in between the periodic inspection, maintenance and certification intervals. These periodic inspections are conducted annually or biennially, and only by an Accredited Service Agent i.e. an external Marine Servicing Company. The reason IMO requires crew to test for contents in-between these is that the "ship sails alone" and "there is no such thing as a small fire at sea"; it must act as its own emergency fire service, differing to a land based asset. In order to create a Safeship® the crew must be in full understanding of the status their safety systems at all times and not just at the times of the inspections – achieved by continuous monitoring.

Gaseous extinguishing systems must be able to actuate, or release their gas, in the event of a fire. Given that the gaseous systems are designed specifically to the individual need of the vessel, and it is known that they leak, then a 5% loss of agent may mean that they would not fully extinguish the fire. These are pressurised systems, so the point of monthly checking for contents is to identify loss of contents through leakage or accidental discharge before they fall below their capacity to extinguish (technically described as: delivering their design concentration). Only having the annual inspection by accredited marine servicing companies is not enough – the crew must take responsibility for its own fire protection and inspecting the gaseous extinguishing systems. But what if there is leakage in between these monthly checks? With fewer, even lower-skilled crew and a greater dependence on autonomous machinery, the dependence on fire systems being checked from shore, let alone on the ship in person, will only become greater. Manual weighing is not only laborious, but also dangerous to the crew conducting the servicing. New technology allows ship owners and crews to help them both the spirit and letter of the regulation and thus know that their vessel is protected in the event of a fire.

Smart technology provides the crew and ship owners with ease of inspection and understanding their extinguishing systems. However, the crew will not be able to refill the gaseous extinguishing system, and instead must rely on notifying the marine servicing company when they arrive at a port, despite the fact that they may only be at the port for a very short amount of time. Due to time pressures, the risk of not being able to find a contractor in time to fill the cylinders in the event of leakage is one that could jeopardise the safety of the entire ship when it is time to set sail. Furthermore, it is well known that vessels are kept at the dock for a minimum amount of time, which reduces time for repairs and thus efficiency without compromise of safety is key. Continuously monitoring the cylinders with ultrasonic sensors that utilises IoT can avoid this, because the network contribution. Using IoT enables the advance notification of the crew and shore based services whilst at sea. Therefore, preparations to address the issues can be made prior to docking to ensure the issues are resolved given the minimum time they have. By having the ability to understand the contents' level of their gaseous extinguishing systems whilst at sea, the crew are also able to comply and exceed the IMO SOLAS FSS and ISO 14520 regulations, whilst ensuring the Safeship®.

2.4 WHAT HAPPENS IF THE REGULATIONS ARE NOT MET?

Without the proper implementation of continuous monitoring, aided by IoT, the Safeship® cannot be achieved and the risk of fire at sea poses continuous threat to the crew, vessel and cargo. Although the implementation of the IMO SOLAS FSS and ISO 14520 codes with continuous monitoring and IoT could drive up costs of purchasing and installation, the long-term savings and benefits far outweighs the initial cost. On quick turnaround trade routes with repairs that are constantly delayed, safety quite often takes second place to meeting their schedule. Maritime companies do not want to lose time or money, and therefore push their personnel to make do, often at the price of safety, or regulatory compliance. Yet this haphazard approach can have severe consequences by endangering lives. If gaseous extinguishing systems were at their full concentration at all times, then in the event of a fire, they would be able to extinguish the fire. However, there are numerous accounts of actuation failures, and some are included below:

2.4 (a) MSC Flaminia, 2012

A clear of example of where actuation has failed to provide safety to the vessel is MSC Flaminia. In July 2012, the container ship was exposed to an uncontrollable fire which tragically lead to three fatalities and two severely injured crew members, as well as dire damage to the ship structure and its cargo. In this example, the actuation of the CO2 system failed when it actuated without instruction in the engine room, although the discharge was intended for cargo hold 4, which turned off the auxiliary boiler and auxiliary fan for the main engine. This led to an out of control fire which required three salvage tugs to deal with the effects of the explosions and fire. However, the extent of the fire meant that the salvage teams could not enter the vessel for 4 days. Cargo areas 3-7 in the ship were significantly damaged and the ships structure was weakened, requiring replacement. Under the pressures, the hatch covers lost their integrity and bulkheads were severely damaged which led to water ingress in all the cargo. [1] The ruling from this event has stated that $280 million of liability will be shared as a result of the incident [2].

2.4 (b) Barzan, 2015

On September 2015, a fire was detected inside one of the cargo holds of Barzan, a Maltese registered container ship. The fixed CO2 system was used but due to a number of leaks in the CO2 line, the required amount of gas did not reach the cargo hold to be effective to smother the fire. The starboard fire main line then developed a large leak at a joint in the under deck passage way and had to be isolated. This restricted the fire-fighting efforts to only the port side, and rendered the starboard side water drenching system unusable. The safety investigation concluded that although the CO2 system and fire mains had been tested satisfactorily prior to the vessel's delivery in May 2015, the quality of the workmanship had contributed to the subsequent failure of both systems [3].

2.4 (c) CCNI Arauco, 2016

A major fire broke out in an after hold of the container ship *CCNI Arauco* at Hamburg's Burchardkai terminal in September 2016. A fire department spokesman said that over 100 emergency services personnel, four fire engines and two fire boats were on site to fight the blaze on the day of the incident. Early efforts focused on cooling the hull to prevent structural damage. An initial attempt to smother the fire with CO2 was not effective, and firefighting operations continued through night. The depth of the fire's location within the hold added complexity to the response. Repeated CO2 discharges from the ship's own fixed firefighting system were not sufficient to halt the blaze in the *Arauco's* hold, and a major shore-based intervention was required instead [4].

Therefore, an increased awareness of the regulations and need for maintenance of these systems promotes a safer vessel, achieved by Safeship® technologies.

3. WATERTIGHT INTEGRITY

3.1 WHAT ARE SEALS AND WHY DO THEY LEAK?

As the main cause of vessel loss is sinking, the maintenance, testing and monitoring of watertight hatches, doors and multiple cable transits on vessels is essential. A watertight hatch cover is designed to prevent the passage of water in either direction under a head of water for which the surrounding structure is designed. Many mariners may think hatches are robust, monolithic structures, thereby failing to appreciate the small tolerances on panel alignment and gasket compression. It is better to think of hatches as complex, finely-made structures, to be handled with care. All types of seals, experience dynamic stresses as part of their operational lifetime. For example, 4mm wear on the steel-to-steel contact is sufficient to damage rubber sealing gaskets beyond repair; 5mm sag along the cross-joint can cause a large gap between the compression bar and gasket. The importance of continually maintaining seal integrity should take a more prominent position in ship maintenance scheduling.

3.2 WHAT ARE THE REGULATIONS THAT GOVERN WATERTIGHT INTEGRITY?

The existing regulations that state the need to ensue watertight integrity, are also part of the IMO SOLAS Code.

IMO SOLAS Reg II-1/13-1.1 – Openings in Watertight Bulkheads and Internal Decks in Cargo Ships - "Where penetrations of watertight bulkheads and internal decks are necessary for access, piping, ventilation, electrical cables, etc., arrangements are to be made to maintain the watertight integrity."

IMO SOLAS Reg II-1/21.3 – Periodical Operation and Inspection of Watertight Doors, etc., in Passenger Ships - "The watertight doors and all mechanisms and indicators connected therewith[sic] shall be periodically inspected at sea at least once a week."

3.3 WHY DOES WATERTIGHT INTEGRITY ALSO NEED CONTINUOUS MONITORING?

The future of watertight integrity testing is with continuous monitoring. A lack of proper servicing of seals can lead to deterioration which endanger the lives of the crew, vessel and cargo. The large issue here is that ships are only tested before and after one or perhaps several journeys; yet a leak could occur at any point in between testing and continue unnoticed until the next inspection. A vessel generates its leak sites due to load states, sea states, wind states, and dynamic movement. The severity is amplified within a vessel structure constantly changing by varying sea, wind, load states, cargo types and dynamic stresses. There is a great deal of bending and deformation that naturally occurs in ships during travel. It was found that a comprehensive, autonomous continuous monitoring system for the watertight integrity of a ship's cargo hatches, weathertight doors and other seals is possible to be developed. One that is capable of automatically detecting emerging leak sites, alerting officers and crew of the location and severity of the leak site and logging all data for future review. The developments in continuous monitoring technology being undertaken by Coltraco Ultrasonics will drive the industry towards ensuring that watertight integrity is never left to chance.

3.4 WHAT HAPPENS IF THE REGULATIONS ARE NOT MET?

Just as with gaseous extinguishing systems, without the proper implementation of continuous monitoring, aided by IoT, the Safeship® cannot be achieved and the risk of water ingress poses a threat to the vessel, crew and cargo.

The case study of the Emma Maersk exemplifies the danger of improper servicing. A severe leakage occurred on the container ship in February 2013 when it was loaded with 14,000 containers. The leakage was caused by the mechanical break-down of a stern thruster, creating the shaft tunnel to flood, as well as leading to severe ingress of water in the aft part. This led to flooding of the main engine room. This was caused by non-effective cable penetration sealings: in a sudden blast, four cable penetration sealings in the watertight bulkhead gave way to the water pressure followed by a massive ingress of seawater. Shortly after this, the other three cable penetration sealings also failed, resulting in an even larger ingress of water into the engine room. This led to approximately USD 45 million worth of damages and towage cost [5].

The NTSB (National Transportation Safety Board) have recently investigated the cause of the 2015 *El Faro* disaster. SS *El Faro* was a United States-flagged, combination roll-on/roll-off and lift-on/lift-off cargo ship crewed by U.S. merchant mariners. All 33 crew members tragically died in the sinking, when *El Faro* sailed from Jacksonville into Hurricane Joaquin, while heading to Puerto Rico. The wreckage was discovered more than 15,000 feet below the sea surface, Northeast of Acklins and Crooked Island, Bahamas. The NTSB have concluded that gaps in safety management contributed to the sinking of the *El Faro*. One of the significant issues was "*poor watertight integrity which allowed seawater into the ship*" stating that this accident may have been avoided if "*crew had more information about the status of the hatches*". The tragedy of the El Faro has exemplified why it is crucial for the watertight integrity of vessels to be upheld [6].

4. CONTINUOUS MONITORING

4.1 INTRODUCING IoT TO TECHNOLOGY

Vessels require constant monitoring for them to become stable at sea, both of their gaseous extinguishing systems and of their watertight integrity systems. Current technologies in the Internet of Things allows this to be viably and innovatively achieved.

The International Telecommunication Union (ITU) defines the Internet of Things as "a global infrastructure for the Information Society, enabling advanced services by interconnecting (physical and virtual) things based on, existing and evolving, interoperable information and communication technologies [7]. It is simply a method of connecting devices from the general household appliance through to sophisticated computers onto the existing internet infrastructure allowing continuous communication and control wirelessly.

To enable such connectivity, there already exists various sources which have defined the importance of IoT architecture, elements and protocols involved and provide an explanation to their various functions [8] [9].

Such connectivity enabled through IoT is gradually being seen as a growing cause of concern particularly in the data privacy and security realm of things. Without neglecting the need to protect and preserve privacy, this section aims to showcase the various opportunities IoT brings to improving safety on ships through the implementation of its technology onboard existing monitoring systems.

4.2 ADVANCEMENT OF THE WIRELESS SENSOR NETWORK (WSN)

One of the main attractions of IoT when coupled with remote sensing is its capability to enhance the value of data collection by providing a pathway for the data to be disseminated instantaneously away from the sensors and into various other processing platforms and monitoring software. Recent technological advances in low power integrated circuits and wireless communication enable the construction of a sensor network consisting of a large number of intelligent sensors; this enables the collection, processing, analysis and dissemination of valuable information in an efficient, low cost and low power manner [10].

This advancement in technology has led to the introduction of the Wireless Sensor Network (WSN) concept. This generally consists of a gateway that can communicate with a number of wireless sensors which are call nodes via a radio link. Data is then collected at the wireless sensor node, compressed and transmitted to the gateway directly or if required, uses other wireless sensor nodes to forward data to the gateway [11]. The hardware involved within a wireless sensor node can generally be summarised in the block diagram below:

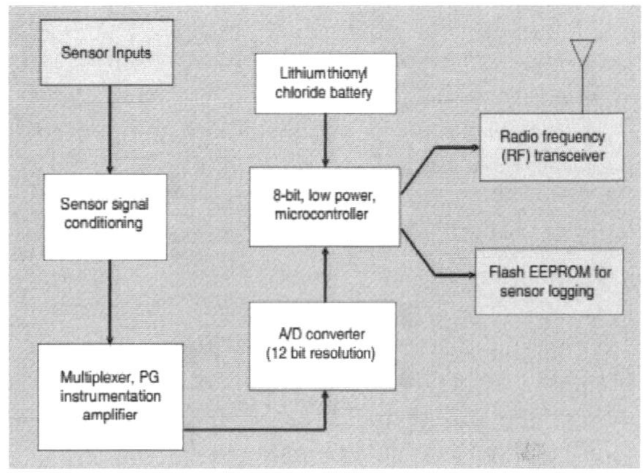

Figure 1: Wireless sensor node block diagram illustrating how a sensor input gets transmitted wirelessly [12].

The second part is the gateway involved in the WSN which is the hardware used to process incoming data sent from the wireless nodes and connects them onto the Internet or Local Area Network to be accessible. The wireless protocol used is highly dependent on application but is commonly based on IEEE 802.11 standards or more widely referred to as Wi-Fi. With the growth and advancement of IoT, the development of WSN is becoming more widespread and cost effective.

4.3 MONITORING FIRE SUPPRESSION SYSTEMS

To prevent fires, regulations require installation of a fixed fire protection system and annual maintenance by a licensed servicing company to check its condition. Following from the above it is now possible to go above and beyond the regulation requirement of annual maintenance which may not be sufficient to cover the remaining periods when the vessel is out at sea alone. The idea of continuous monitoring of fire suppression systems has been made possible and is done by non-invasive ultrasonic liquid level monitoring of the contents within the fire suppression cylinders. This ensures the quantity of agent required to extinguish a fire is always at its designed concentration to guarantee an efficacious fire protection in the event of a fire breaking out in the vessel.

Liquid level monitoring is done using compact, low-cost ultrasonic sensors that are linked to a central control module that broadcasts the liquid level information over WSN allowing dissemination of information to the appropriate parties. The ultrasonic sensors are attached to the walls of the cylinders and propagates ultrasound pulses which travel across the cylinder and returns to its starting position. As the ultrasound pulses travel across the cylinder, the contents within the cylinder will absorb the signal energy in proportions relative to their phases. This ensures that the difference in returned signal above and below the liquid level is always sufficiently large, allowing the liquid level to be identified each time. By fixing the sensor onto the cylinder permanently, any changes in the contents' level will be identified immediately. Research conducted in the labs have proven accuracies up to 0.1 mm when tested with the commonly known fire suppressing agent cylinders. This proprietary system by Coltraco Ultrasonics is the Permalevel® Multiplex and is capable of using the latest advancement in IoT to collect and disseminate sensor information and forwards them onto the control panel situated on the vessel's bridge to be integrated as part of the vessel's overall monitoring system.

4.4 MONITORING WATERTIGHT INTEGRITY

To prevent sinking, regulations outline the requirements to ensure watertight integrity checks are done before the vessel sets sail. There is currently no emphasis on periodic watertight integrity checks other than to ensure watertight doors are closed. There are also reasons to believe that this requirement can too be exceeded by implementing the continuous monitoring of watertight compartments such as hatch covers for cargo holds and watertight doors for passenger spaces. This is now made possible by implementing robust ultrasonic transmitters and receivers both inside and outside the monitored compartment respectively. As soon as apertures start to form in the seals of these structures, the ultrasonic signal escapes the compartments and is received by the ultrasonic receiver on the other end. Above a certain threshold, this escaped signal represents the size of the leak that has developed.

Using a portable ultrasonic watertight integrity tester, Coltraco Ultrasonics' Portascanner® functions on similar principles, detecting the smallest leak size 0.06 mm in diameter which is the most mathematically accurate in the world +/-0.02mm. As an example of how a leak will compromise the stability of a vessel, a leak diameter of one mm under just one metres depth of water below sea level will correspond to a leak of 12,500 millilitres per hour. This means up to 911 cans of soft drinks worth of water in a day. This highlights the importance of watertight integrity.

Despite the increasing uptake of portable ultrasonic watertight integrity testers in the market, the application of ultrasonic technology to continuously monitor the watertight integrity in vessel compartments has never been done before. Companies like Coltraco Ultrasonics are integrating their expertise in ultrasonic sensors from their portable ultrasonic watertight integrity tester, the Portascanner®, into a full continuous monitoring system, the Permascanner® able to rely on WSN developments to transmit information on developing leaks across the vessel into the bridge and beyond.

4.5 EXAMPLE OF WSN AS EVIDENCED BY THE PERMALEVEL® MULTIPLEX AND ITS BENEFITS

Given that both the Permalevel® and Permascanner®'s primary functions are to protect a vessel from its two largest causes of loss, it is beneficial and more efficient to link their sensor information to a network accessible by relevant parties via WSN.

Building upon the previous' section description of how a WSN functions, an example of how the anticipated pathway of data will flow in the Permalevel® system is illustrated below:

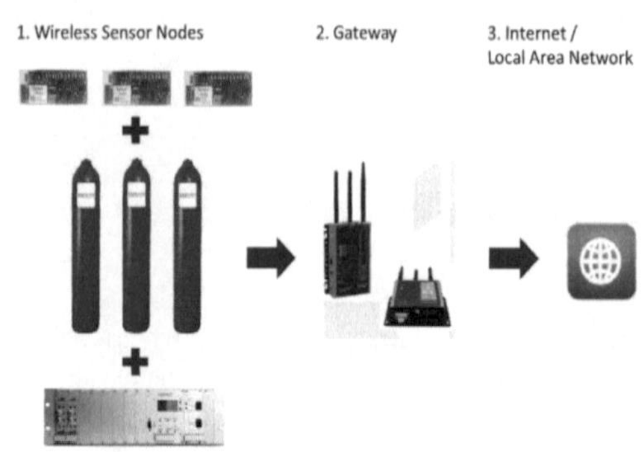

The first section of the Permalevel® system as part of the WSN is its sensor inputs assigned to a wireless node each. As described in the previous section of how liquid level is detected using an ultrasonic sensor, each sensor captures the information into their respective wireless node which then forwards the information towards the gateway as shown in the second section. These wireless nodes communicate with each other and the gateway following the protocol IEEE 802.15.4, 2.4 GHz with a self-healing SNAP Mesh network to ensure low-power and high reliability. The gateway then continues to forward these information via the standard wireless protocol IEEE 802.11, 2.4 GHz or built-in Ethernet onto the online network. Once the information is forwarded online, the information becomes accessible anywhere with a computer, an Internet connection and the appropriate software configuration.

This information is logged and stored, and used for important functions such as triggering an alarm whenever the sensors are being tampered with or when the cylinders' contents fall below the desired level. This helps ensure that the vessel's fire suppression systems are always maintained at the full concentration levels required to extinguish fires. Full visibility of the vessel's fire suppression system is also granted not only to the vessel's captain on the bridge, but also to shore-based operators monitoring the movements and various other vessel conditions.

Because information collection from the WSN are typically logged and recorded, it can also be used to prove due diligence for insurance companies in the event of a damage claim due to fire for example. This protects a vessel owner's reputation, as well as the reputation of those manning the ship if the appropriate steps were taken to ensure the fire suppression system is well maintained.

Collected data sets can also identify weak points that typically go unnoticed in an inspection. For example, there has been a reported incident which is noted by the Maritime Executive in that the U.S. Coast Guard marine inspectors actually identified a crack in the vessel's CO2 extinguishing system hose during their periodic inspection that may have otherwise gone unnoticed despite the servicing and maintenance already done by a licensed marine servicing company [12]. If this crack in the hose had caused the CO2 system to leak, monitoring the CO2 cylinders' contents would quickly identify the issue allowing remedial action, alerting the relevant personnel in real time through the sensors' wireless connectivity.

4.6 AN IoT'S CONTRIBUTION TO IMPROVING EXISTING REGULATIONS

Prior to the advancement of WSN, there existed no simple and cost-effective method to transmit information from sensor to sensor and to the vessel's bridge and onwards other than a physical wired connection. WSN overcomes this difficulty by providing a reliable pathway for the dissemination of sensor information that can be viewed from a variety of sources for different purposes anytime, anywhere.

An assumption of why regulations tend to remain stagnant despite the ever-growing safety concerns which pinpoint their insufficiency is in the fact that there exists no better solution and technology to address the issue. One of reasons why annual or biennial inspections are the way they are is the perceived lack of tools available to make more frequent and viable inspections. Companies such as Coltraco Ultrasonics are one of the few championing innovation and demonstrating its viability to successfully address issues with the growth of IoT.

As the marine industry continues to face pressure regarding the ever-growing safety concerns yet unaddressed by a regulation change, countries such as Singapore are beginning to accept the advancement of IoT into the Singapore Registry of Ships in order to stay ahead. According to the Maritime and Port Authority of Singapore (MPA)'s chief executive Andrew Tan, IoT, digitalisation and new technologies such as blockchain and smart drones are changing the way things work. To stay ahead, the Singapore Registry of Ships needs to embrace these technologies to offer value-added service to its customer [13]. This prompt action may well be one of the many important ones which may influence the safety of many in an industry whose safety is mainly governed by the regulations.

5. CONCLUSION

The Smart Ship can be a Safe Ship by implementing new technology such as that developed by Coltraco Ultrasonics. The neglect of continuous monitoring of the fundamental protection provided by the gaseous extinguishing systems and watertight integrity imperils the lives of ship's crews and occupants and at the risk of crippling financial and reputational loss to the tanker owner. Continuously monitoring gaseous extinguishing systems and watertight integrity with the aid of IoT raises awareness of the safety critical need to implement and exceed regulations, through continuous inspection.

Coltraco Ultrasonics understand that through the developments and current emphasis on IoT we are closer to reaching the Smart Ship which is bringing greater awareness for the shipping industry to improve. Due to this, future regulations could be strengthened by the inclusion of constant monitoring and IoT developments to allow for the better implementation of the Smart Ship which will be a Safe Ship. In conclusion, this paper argues that IoT and big data offer great benefits to enable implementation and potential change in the regulations.

6. DISCLAIMER

The opinions expressed are those of the author.

7. ACKNOWLEDGEMENTS

The author would like to thank those at Coltraco Ultrasonics for in-house research and to Durham University Physics Department and the Centre of Industrial Ultrasonics at the Department of Physics and Engineering at Warwick University who undertook research for Coltraco Ultrasonics. Thank you also to industry professionals who have taken the time to review some of this paper prior to publishing.

8. REFERENCES

1. Bundesstelle fur Seeunfalluntersuchung., 2014. *Fire and explosion on board the MSC Flaminia On 14 July 2012 in the Atlantic and the ensuing the events,* s.l.: http://www.bsu-bund.de/SharedDocs/pdf/EN/Investigation_Report/2014/Investigation_Report_255_12.pdf?__blob=publicationFile. [Accessed 06 01 2018].

2. TradeWinds, 2017. *Ruling Sets Stage for USD $280m liability to be shared.* [Online] Available at: http://www.tradewindsnews.com/liner/1386593/ruling-sets-stage-for-usd-280m-msc-flaminia-liability-to-be-shared [Accessed 06 01 2018].

3. MARINE SAFETY INVESTIGATION REPORT, 2016. *Safety investigation into the fire inside a cargo hold on board the Maltese registered container ship Barzan.* [Online] Available at: https://mtip.gov.mt/en/Document%20Repository/MSIU%20Documents/Investigations%202015/MV%20Barzan_Final%20Safety%20Investigation%20Report.pdf [Accessed 06 01 2018].

4. FleetMon, 2016. *Container ship CCNI ARAUCO fire: Sep 04 afternoon Update.* [Online] Available at: https://www.fleetmon.com/maritime-news/2016/14647/mega-container-ship-ccni-arauco-explosion-and-fire/ [Accessed 06 01 2018].

5. Danish Maritime Accident Investigation Board, 2013. *MARINE ACCIDENT REPORT EMMA MÆRSK Flooding of engine room on 1 February 2013.* [Online] Available at: http://www.dmaib.com/Ulykkesrapporter/Flooding%20of%20engine%20room%20-%20EMMA%20M%C3%86RSK%20on%201%20February%202013.pdf [Accessed 06 01 2018].

6. National Transportation Safety Board, 2017. *Captain's Decisions, Shipping Company's Poor Safety Oversight Led to Sinking, NTSB Says.* [Online] Available at: https://www.ntsb.gov/news/press-releases/Pages/PR20171212.aspx [Accessed 06 01 2018].

7. Internet of Things Global Standards Initiative, ITU, 2012. https://www.itu.int/en/ITU-T/gsi/iot/Pages/default.aspx. [URL last assessed 5th January 2018.]

8. I.F. Akyildiz, W. Su, Y. Sankarasubramaniam, E. Cayirci. Wireless sensor networks: a survey, Computer Networks, 38 (2002), pp. 393-422.

9. Al-Fuqaha A., Guizani M., Mohammadi M., Aledhari M., Ayyash M., Internet of Things: A Survey on Enabling Technologies, Protocols and Applications. IEEE Communication Surveys & Tutorials, Vol. 177, No. 4, Fourth Quarter 2015. pp. 2347 – 2376.

10. Atzori L., Iera A., Morabito G., The Internet of Things: A survey. Computer Networks, 54 (2010), pp. 2787 – 2805.

11. Townsend, C. and Arms, S. (2005). Wireless Sensor Networks: Principles and Applications. In: J. Wilson, ed. 1, Sensor Technology Handbook. Oxford: Elsevier, pp.439-449.

12. Marex, Coast Guard Warns on Fixed CO2 Fire Extinguishers, The Maritime Executive, 29th December 2017. https://www.maritime-executive.com/article/coast-guard-warns-on-fixed-co2-fire-extinguishers. [URL last assessed 10th January 2018.]

13. Chambers S., MPA Using Drones for Ship Surveys., Splash 24/7, 27th November 2017. http://splash247.com/mpa-using-drones-ship-surveys/ URL last assessed 10th January 2018

14. Hunter, C.S.P (2017). The Ungoverned Space. The Journal for Maritime Procurement and Supply Chain management, (7), pp.36-38.

15. Hunter, C.S.P (2016). Ultrasonic technology for reliable ship maintenance. Oil Review Middle East, (8), pp.31-32.

9. AUTHORS BIOGRAPHY

Dr Carl Stephen Patrick Hunter *BA(Dunelm) Hon DSc FRINA FIMarEST MRAeS is CEO and Managing Director of Coltraco Ultrasonics, a British designer and manufacturer of portable and fixed monitoring systems for the naval, shipping, offshore, energy and fire sectors. He is a former GreenJacket Officer in the British Army and a Graduate and Honorary Doctor of Science from the University of Durham, a Fellow of the Royal Institution of Naval Architects, a Fellow of the Institute of Marine Engineers and Member of the Royal Aeronautical Society, Royal Institute for International Affairs, Royal Society of Asian Affairs and Royal United Services Institute. Dr. Hunter is a Council Member of the British Naval Equipment Association, a member of the Export Council of the Fire Industry Association(FIA) and the FIA's Working Group Gases which is the Lead UK Fire delegate organisation to British Standards and BSI the lead fire delegate to the International Standards Organisation in Gaseous Extinguishing Systems and an active member of the Society of Maritime Industries the lead British marine organisation and voice of the marine industry in the UK. His interests are in Physics, International Business, the Constitution, South Asia, International & Strategic Affairs, Defence, the British Commonwealth and bringing UK Scientists into business. Dr Carl lives in London and Somerset in England, travels extensively in Asia, Europe and the Middle East, has been married for 27 years to Dorothy and a proud Father of four Children.*

Adrian Saw*, Bsc Dunelm, Research Manager, Coltraco Ultrasonics, UK.*

Chloe Kay*, BA Dunelm, Marketing & Communications Assistant, Coltraco Ultrasonics, UK.*

Clare Mary Elizabeth Hunter *BA Dunelm, Head of Marketing & Communciations, Coltraco Ultrasonics, UK.*

MARITIME ENERGY MANAGEMENT SYSTEM: FROM AUTOMATIC DATA LOGGING AND DIAGNOSTICS TO OPERATIONS SUPPORT TOOL

S Y Jeong, D H Jeong, J M Park, J H Park, B R Kim, and **K S Ahn,**
Hyundai Heavy Industries Co., Ltd., Republic of Korea

SUMMARY

Energy management in the maritime industry is constantly being required by economical reasons and in conformity with new legislative actions taken by the International Maritime Organization (IMO) and the European Union (EU). In response, various performance monitoring methodologies and data collection practices have been examined by different stakeholders. Despite several advancements in ship operations and its related technologies, their adoptions in the shipping industry stay small. This slow uptake is considered due to many different barriers such as data analysis problems, misreported data, and feedback problems, etc. This study presents a configuration and functional descriptions of an energy management system (EMS) developed to resolve the problems mentioned above. The system automatically gathers data, analyse them, and support users in their planning voyages based on the analysis results. It is also expected to help assess the ship performance properly and implement sustainable fleet management.

1. INTRODUCTION

Generally, bunker fuel costs account for 40-50% of the operating costs in maritime shipping. International oil price has increased rapidly from 2000 to 2008 of bankruptcy of Lehman Brothers. In this period, several energy monitoring or management systems were developed and attracted shipping industry. Although the oil price plummeted in July and August 2014, efforts to reduce management costs continued to overcome deteriorating management environment of shipping companies.

On the other hand, interest in energy monitoring of vessels is steadily increasing as international regulations such as International Maritime Organization (IMO) and European Union (EU) strengthen regulations on CO_2 emissions. Since 2003, IMO has begun to discuss specific details of GHG (Green House Gas) regulations [1]. IMO seeks to regulate greenhouse gas emissions from ships through three methods: regulation in new building vessels, regulation in ship operations, and market-based measures. The EEDI (Energy Efficiency Design Index) [2] for new building vessels and the SEEMP (Ship Energy Efficiency Management Plan) [3] for existing ships are included in the MARPOL Convention and will be enforced from 2013. Meanwhile, the EU submitted a law on execution of MRV (Monitoring, Reporting, Verification) to the European Parliament. It requires all ships over 5,000 GT to enter and exit EU ports to measure CO_2 emissions and other climate-related information every year and submit to EC (EU Council) with third-party verification. IMO SEEMP or EU MRV requires common automated data collection and reporting. This is because it is necessary to prepare data acquisition and collection methods for data verification as well as data recording.

2. CONCEPTUAL DESIGN

2.1 REQUIREMENT AND APPROACH TO SOLVE

MARPOL Annex VI is regulation regarding the prevention of air pollution from ships. It was revised at IMO Marine Environment Protection Committee (MEPC)'s 70th meeting in October 2016. Revised version considers recording and reporting the fuel consumption data. MEPC guidelines for SEEMP provide a framework for the following four steps; planning, implementation, monitoring, and self-evaluation & improvement. The design for the EMS proposed in this paper considers supporting the energy management through a series of four actions.

MEPC also represents best practices for fuel-efficient operation of ships in guidelines. It includes improved voyage planning, speed optimization, optimum trim, and so on. We also adopted various measures of the 'fuel-efficient operations' and 'optimized ship handling' mentioned in the MEPC as tools for efficiency improvement.

A variety of EMSs have been developed and released on the market. Shipping companies also have been considering the adoption of EMS in an aspect of international compliance and market demands as described in the previous chapter. In practice, however, they are not widely applied as a best practice even between 2012 and 2014 despite the high fuel prices. Rene and Hannes [4] studied the reasons of low rate of system adoption. They interviewed 55 professionals in 34 shipping organizations in Denmark. Fundamental problems said to interfere with energy management and fuel-saving efforts are listed in Table 1. Five main evident problems were summarized as lack of transparency, data collection, data misreporting, data analysis, and absence of on-shore feedback regarding their study. They see that EMS can move to best practice by solving these challenges.

Approaches in Table 1 can be a solution for these problems. In the following sections, detailed explanation s will be represented in this point of view.

Table 1: Fundamental Problems of EMS Application and Approach for Solution

Fundamental problems	Approach
Lack of transparency	Initial drawings & shop test results Auto data gathering and storage Apply virtual sensors
Data collection challenges	Use data platform connected with AMS, VDR
Incentives for data misreporting	Install flow meter to M/E and G/E
Data analysis problems	Normalize data considering weather condition
Feedback problems	Transfer all the data and summarize them

2.1 (a) LACK OF TRANSPARENCY

There are five causes about energy management data sets to obstruct seeing the results of fuel-saving efforts [4]. They are lack of baselines, lack of long time-series data, lack of real-time data sets, lack of sub-metering on-board, and noise in data sets. They can be supplemented by comparison of data by using ship design data, shop test, model test or sea trial result data. For example, when a flow meter value is biased or missed, the data can be estimated using the log speed, GPS speed, or main engine power. System combines all the data related to fuel flow, and it will help indicate clearer sensor value. It is the concept called virtual sensor or sensor fusion [5].

2.1 (b) DATA COLLECTION CHALLENGES

There are two fundamentally different approaches to gather energy consumption data. One is auto-logging, and the other one is manual logging system [4]. In the case of manual data input mode of EMS, it is difficult to acquire a long time-series data. The proposed system of this paper basically requires the automatic data logging.
However, some are worried that on-board crews can lose involvement in energy saving efforts if they are no longer responsible for data collection. In addition, a large number of vessels are not equipped with enough online sensors. So the essential data to be gathered should be predetermined. It needs to be connected online through data platform which is gathering data from AMS (Alarm Monitoring System), VDR (Voyage Data Recorder), and individual Data Acquisition Unit (DAU). For sustained focus on energy management system, manual input data such as changing the status of voyage are selected and system prompts officers in navigation and engine department to enter data.

2.1 (c) INCENTIVES FOR DATA MISREPORTING

Data misreporting is about a moral hazard problem. It is caused by the lack of vessel flow meters. There are three examples of moral hazard problems [4]. One is over-reporting fuel consumption for selling on the black market. Crews store the excess fuel, and then sell it under the counter. Another is making false entries about the amount of daily fuel oil consumption for avoiding the risk of legal action from port state control (PSC). Actual fuel deliveries often turn out to be short of the quantities specified in Bunker Delivery Notes (BDN). To reconcile the discrepancy, On-board daily fuel consumption is over-reported. The other one is to misreport for charter party. In case of time charter, charterer pays for fuel. So the limit of fuel consumption is specified. To avoid claim from charterers, ship owner encourage crews to underreport fuel consumption. For these reasons, many experts from shipping companies agree on the need for fuel flow meter on board. However, large amount of vessels in the world are not equipped with the indicators. In order to overcome the issues, flow meters connected to M/E and G/E is basically supplied at new building stage with this EMS. This will record the actual fuel usage metrics so that it will be possible to respond appropriately to the third party's data validation issues.

2.1 (d) DATA ANALYSIS PROBLEMS

Normalizing process of raw data is required to use general data analysis method. However, it is still a difficult problem especially considering the impact of external environment. Measured data such as delivered power values shall be corrected considering weather effects. In this article, correction for weather-caused factors is calculated according to the procedure of international standards. ISO 15016: 2002 [6] and ISO 15016: 2015 [7] are applied to ship speed performance analysis, and ISO 19030 [8] is applied to measurement of changes in hull and propeller performance.

2.1 (e) FEEDBACK PROBLEMS

Crewmembers and shore staffs should be in continuity and communicate with their feedbacks. Lack of feedback reduces their awareness of energy management. For the connectivity, communication between onboard and onshore via IP-based satellite communication network environment can be considered. In other words, onshore managers can review the operational status, voyage history of each vessel, voyage plans and analysis results. It is also possible to compare performance indicators for vessels in their fleet. This means the onshore manager, who has a variety of information, will be able to present a guide for each vessel or control the fleet. Conversely, vessels may request support from the onshore technical department for maintenance.

2.2 SYSTEM CONFIGURATION

Vessel information can be grouped under two categories, navigational information and engineering information. The formal one is information such as GPS time, location, speed, etc. ECDIS (Electronic Chart Display and Information System) and VDR are the vessel-mounted device related to this information. The latter one is information related to the device the engineer is handling, e.g., M/E and G/E information. This kind of information is gathered and managed by AMS. Therefore, EMS in this paper is connected to ECDIS, VDR and AMS to collect ship data.

EMS server which is interfaced with vessel's statutory devices manages data and processes analysis functions. Users can access the server using a client PC to use the services. Data server has four LAN ports to connect VDR, AMS, client PC in ship network and satellite equipment. Figure 1 is an example of network diagram of EMS. It may vary depending on the network environment in the ship.

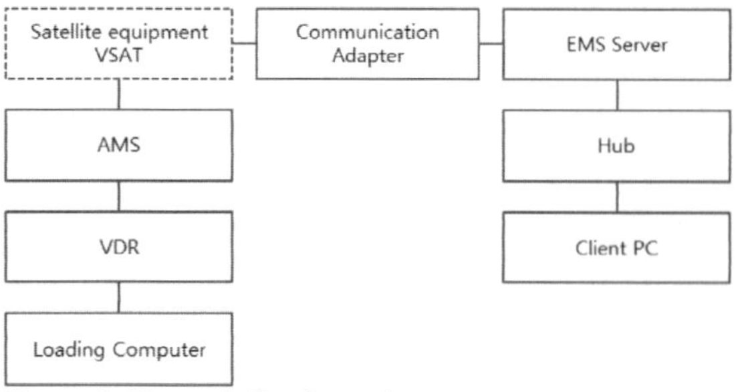

Figure 1: Network Configuration

VDR and AMS are used to collect the sensor data needed for EMS, and the client PC in ship network access to the EMS web services. The satellite equipment is required to connect the on-board platform to the on-shore server. On-shore server can access on-board server. Users can check the screen of a ship in real time. On-board server also transfers the sensor data on a regular cycle.

The system in this paper uses PostgreSQL as a database management system. Data collected from each sensor is stored in four kinds of DB tables according to the time interval as described in Table 2.

Table 2: Storage of collected data

Update period	Duration of storage
1 seconds	2 months
10 seconds.	15 minutes
1 minute.	Permanent
1 day	Permanent

The main functions of the proposed system in this paper are listed in Table 3.

Dashboard provides functions that are monitoring the ship's condition and state in real time. Users can use the sub-functions according to their roles and needs. Voyage monitoring function provides the current voyage state in real time against the plan. Vessel ranking function provides energy management insight through comparison with other ships as well as fuel efficiency, navigational position and speed.

Performance analysis provides in-depth analysis of the vessel's speed performance and visualizes the results. It analyses the actual performance of the ship using normalisation to exclude the influence of the environment. It also shows the operational profile what condition the ship sails. It can give users an insight into the fleet management or design of new ships. In the long term, it can estimate the hull fouling through the analysis on the power and speed performance of the ship. The result can be used as a reference for determining a dry dock schedule.

Voyage planning is directly involved in improving fuel efficiency through navigational plan and a fuel-efficient operation. It recommends optimal speed, path, and trim of the ship based on performance estimated from the ship's unique characteristics. It helps operators to carry out the same transportation mission with less energy.

Machinery management shows operational status and records on various devices such as M/E. G/E, and boiler. This allows crews to check the actual fuel efficiency of the individual equipment. It compares the operational status to a reference curve such as shop test or sea trial result.

On-board report supports the creation of various types of reports on board. By enabling automatic entry of logged data such as moving distance, wind direction and wind speed, it helps reduce the number of repetitive tasks that officers write manually.

A detailed description of the functions and tools described above is provided in the following sections.

Table 3: Menu Hierarchy Diagram

Level 1	Level 2
Dashboard	- Voyage monitoring - Energy flow - ENC (Electronic Navigational Charts) - Fuel management - Vessel ranking
Performance analysis	- Speed performance analysis - Weather effect analysis - Operational profile - Machinery performance - Hull fouling analysis
Voyage planning	- Route optimization - Trim optimization
Machinery management	- Main engine - Generator engine - Boiler
On-board report	- Noon report - Noon report (in port) - Departure report - Arrival report - Voyage report - EU MRV report - IMO DCS report

3. FUNCTION DESCRIPTION

3.1 DASHBOARD

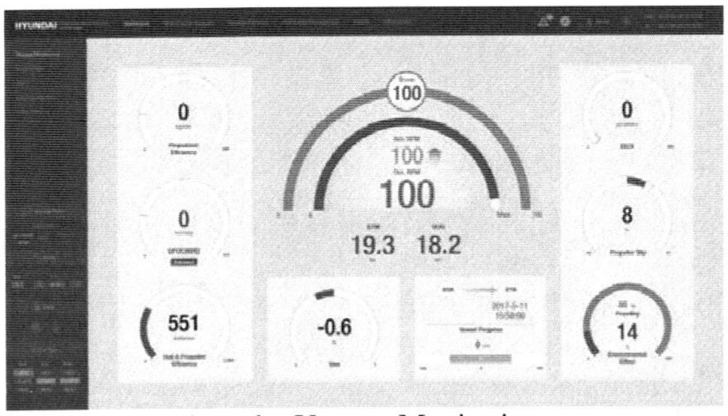

Figure 2: Dashboard – Voyage Monitoring

The dashboard (Fig. 2) provides general information on the ship's operational status and route. The functions provided are composed of the following elements:
- Voyage monitoring
- Energy flow
- ENC (Electronic Navigational Charts)
- Fuel management
- Vessel ranking

Voyage monitoring shows a variety of indicators that shows current voyage conditions at a glance, especially with comparing the ongoing state with the best navigational state. This helps officers to check for changes in excess fuel use or fuel efficiency indicators versus planned.
Energy flow is a function of tracing the ship's energy uses. Fuel oil, natural gas, and other energy sources can be diverted to propulsion, electricity, steam, and the energy loss in the process. For the generated electricity, it is possible to measure the amount of power of each load and distribute according to the usage. If additional sensors are not installed, the ELA (Electric Load Analysis) can be used to check the power distribution and consumption of the loads although precise measurements are not possible.
ENC is the electronic navigational chart function that represents the current vessel position and planned route. It does not provide only a simple electronic chart function, but also weather information and weather change patterns can be displayed. Weather forecasts such as wind, wave, and ocean current can be displayed on an interactive animated map.
Fuel management provides displays of the fuel level and schedule manager function for each type of oil (confirmation of the type of oil used by each vessel, time history of oil change, oil bunkering and balance correction).
Finally, **vessel ranking** provides the function to compare the energy efficiency and performance of a ship with that of other vessels in the fleet or compare it with the past record of the vessel. This can be used not only to confirm the relative performance of the ship but also as data for management of the fleet.

3.2 PERFORMANCE ANALYSIS

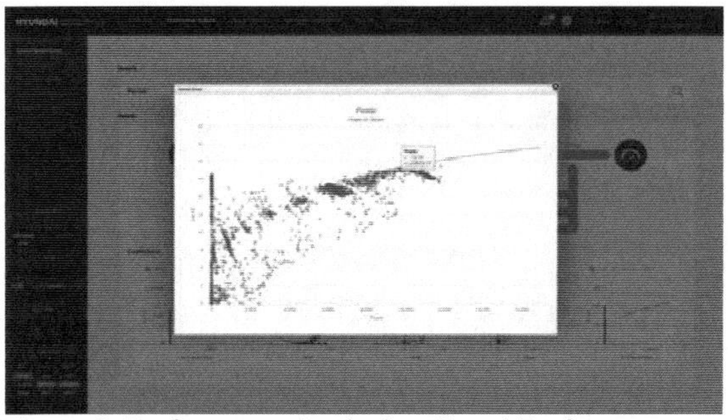

Figure 3: Performance Analysis – Speed Performance

Performance analysis consists of the following functions:
- Speed performance analysis
- Weather effect analysis
- Operational profile
- Machinery performance
- Hull fouling analysis

Speed performance analysis shows the relationship of the main factors that can represent the performance of the ship from the fuel to the speed during the navigation. The four elements defined here are the fuel mass flow of the ship, the power output from the engine using fuel, and the propeller speed (RPM) of the ship directly connected to the engine.
Weather effect analysis defines the parameters of engine horsepower during the voyage of the vessel, and shows the influence of each environmental variable on the voyage. It also shows the results of analysing the speed performance at sea based on ISO 15016: 2002 and ISO 15016: 2015.
Through **operational profile**, it is possible to trace the navigational record of the ship as a histogram. These data are categorized into 'voyage', 'weather', and 'fuel'. 'Voyage' consists of ship's speed, engine horsepower, propeller speed, draft and trim. 'Weather' is about winds and waves encountered. They use weather forecast as basic information, but also display the measurement data of the ship in case of wind. 'Fuel' indicates the amount of fuel used by each piece of equipment.
Machinery performance helps compare several items such as SFOC for each engine load for three pieces of equipment: M/E, G/E and boiler. The references of the comparison are shoptest, sea trial and past voyages. With this function, users can check the performance degradation versus commissioning or the past voyages.
Finally, **hull fouling analysis** shows the trend over time by analysing the influence of Hull fouling on the ship's operation. Two methods for analysing hull fouling in this system are ISO 19030 and our own machine learning techniques.

3.3 Voyage Planning

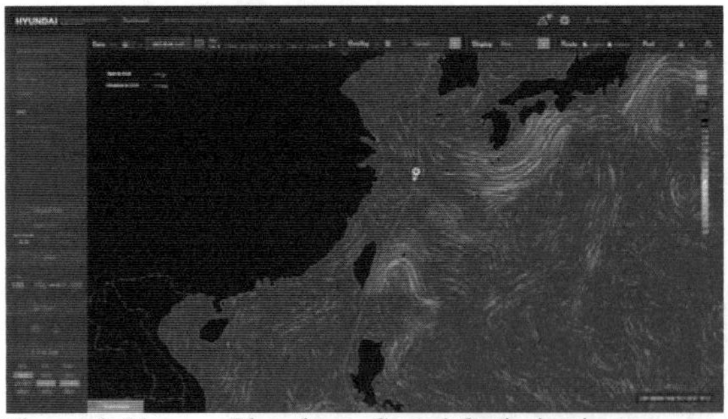
Figure 4: Voyage Planning – Speed Optimization

Voyage planning helps to minimize fuel consumption by analysing the route and trim, and consists of following optimization tools:
- Route optimization
- Trim optimization

Route optimization finds the optimal RPM to minimize fuel consumption and CO2 emissions considering arrival time and weather conditions.
This function allows users to set an optimal voyage plan. To do this, the system in this paper uses the latest weather forecasts, the ship's unique speed performance at sea, and the planned RPM and ETA information that the user initially enters.
When the ship is sailing in the sea, external environmental condition such as wind or wave can affect to the performance of the ship. This kind of disturbance increases the engine horsepower to maintain the engine RPM. It inevitably leads to an increase in fuel consumption. In addition, the speed of the vessel is reduced at the same RPM and additional fuel is used as the navigational time increases when the vessel maintains RPM. Taking this into account, this tool uses the weather forecasts of National Oceanic and Atmospheric Administration (NOAA) to predict the performance of the vessel at sea. Through this prediction, an optimal voyage plan that satisfies the RTA (Required Time of Arrival) and consumes the least amount of fuel can be found.
Next, **trim optimization** suggests an optimal trim in planning voyage conditions such as ship speed, draft, and displacement using CFD (Computational Fluid Dynamics) and model test results.
Even if the ship is in the same displacement condition, the resistance performance may vary depending on the trim condition. Therefore, the resistance performance according to each trim should be calculated or experimented. Optimal trim is recommended in the current voyage planning condition based on this data. Additional ballasting may be required to change the trim of the ship. Therefore, we propose a case in which the horsepower is reduced due to the change of attitude among the cases where the draft increases with the increase of the displacement relative to the initial condition. Users can select the cases that can be realized through actual loading or ballasting among the cases shown in the page.

3.4 Machinery Management

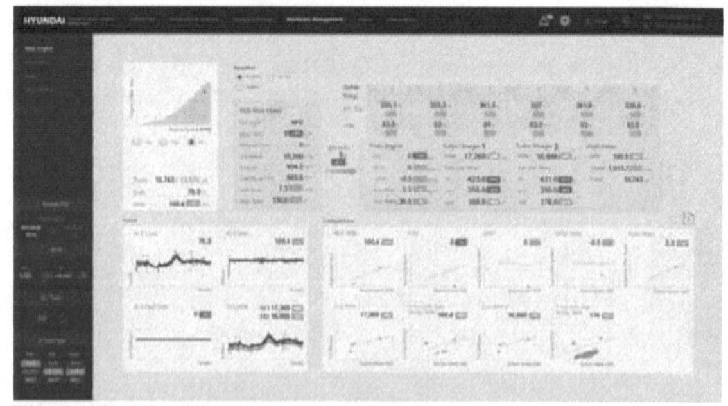
Figure 5: Machinery – Main Engine

Machinery management provides monitoring, logging, diagnostics and analysis functions for the following equipment used in ship:
- Main engine
- Generator engine
- Boiler

While the AMS only displays the current data, EMS logs the data during the life cycle of the ship and provides the change for the last 24 hours. The diagnostic function is comparing and analysing the voyage data with shop test results of the ship. The threshold for the abnormal range can be set to ratio or value. The analytical function generates reports similar to shop test result for the past voyages. Gradient boosting, one of the machine learning algorithms, is used in the machinery performance analysis.

3.5 On-board Report

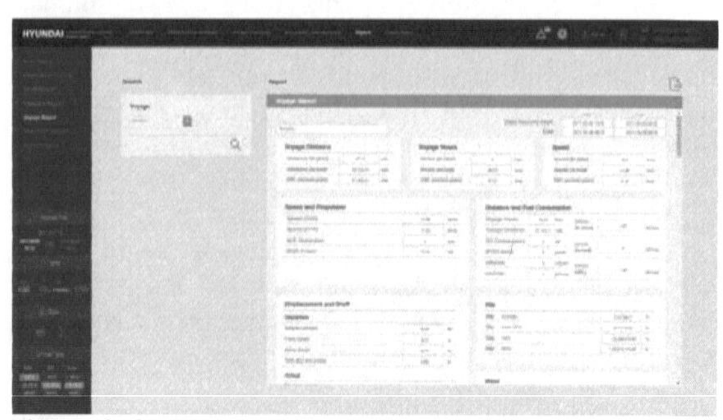
Figure 6: On-board Report – Voyage Report

The on-board report tool will generate reports for various purposes on board the vessel. The report includes the following types:
- Noon report
- Noon report (in port)
- Departure report
- Arrival report
- Voyage report
- EU MRV report

- IMO DCS report

The information automatically generated varies depending on the type and characteristics of the sensors installed on the ship and whether it is connected to the EMS. Reports can be generated as excel files and PDF files.

3.6 Onshore data analysis

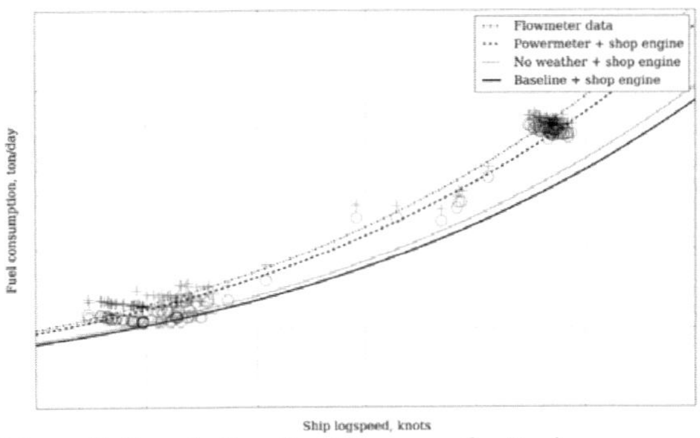

Figure 7: Sample Result – Diagnoses for Fuel Consumption in Service

To analyse and evaluate the performance of the ship, data of sufficiently longer period of time than one voyage is required. Through the EMS designed in this paper, the data collected from the ship are periodically transmitted to the onshore server.
The data collected from the ship have noise due to equipment's specific limitations or external interferences. Therefore, process for refining data is required to generate meaningful data for each analysis purpose. During this process, involvement of expert contributes to ensure the quality of the analysis and data, thus a relatively deep analysis of data over a long period of time is performed on shore. The analysis fundamentally reveals the changes in the performance of the ship and the resulting increase in fuel consumption using the measured speed, horsepower, and shaft speed. The results can be used to define changes in the performance of the ship or evaluate the energy-saving efforts of crews. Vessel's actual operating conditions vary from time to time. However, the reference operating environment is no wind and no waves, at a specific depth and 15°C sea conditions. Therefore, we applied a mathematical model considering ship's physical behaviour, resistance and propulsion performance to normalize ship data. The techniques used are the international standards of ISO 15016 and 19030.
As a result, the amount of excess fuel used by the ship was calculated with respect to the design reference point and estimated by classifying the amount of fuel used by the external environment, hull, propeller, and engine. The results can be expressed in the form shown in Figure 7.

4. CONCLUSION

This paper proposed an approach to overcome the five main categories of obstacles that impede being a best practice for EMS. We sought naval architectural way to solve the problem such as lack of transparency, data collection challenges, incentives for data misreporting, data analysis problems, and feedback problems. In addition, data access/acquisition or satellite access was considered.
We also introduced several services and tools to support on-board officers. Proposed EMS fully supports the process of planning, implementation, monitoring, and self-evaluation & improvement as proposed by SEEMP and MRV. These services were developed based on experience and knowledge of ship operation, naval architectural analysis methods, and system/equipment integration.
Future studies will examine the accuracy of the analysis of data at sea and the reliability of performance prediction model. It is also necessary to review the integration of on-board equipment and the cyber security for implemented system.

5. REFERENCES

1. IMO A.963 (23), IMO Policies and Practices Related to the Reduction of Greenhouse Gas Emission from ships, Dec. 2003.
2. MEPC/Circ.684, Interim Guidelines for voluntary use of the ship energy efficiency operational indicator (EEOI), 17 Aug 2009
3. MEPC/Circ.683, Guidance for the development of a ship energy efficiency management plan (SEEMP), 17 Aug 2009.
4. R. T. Poulsen and H. Johnson, The logic of business vs. the logic of energy management practice: understanding the choices and effects of energy consumption monitoring systems in shipping companies, 2016
5. J. Park, B. Kim, S. Jeong, J. Park, D. Jeong, and K. Ahn, A Comparative Analysis of Ship Speed-Power Performance Based on the Noon Reports and Recorded Sensor Data: Overcoming Sensor Issues in Oceans Conference, 2017
6. ISO15016:2002, Guidelines for the assessment of speed and power performance by analysis of speed trial data, 2002
7. ISO15016:2015, Guidelines for the assessment of speed and power performance by analysis of speed trial data, 2015
8. ISO19030, Measurement of changes in hull and propeller performance, 2016.

6. AUTHORS BIOGRAPHY

Seyong Jeong, Donghoon Jeong, Jinmo Park, Boram Kim, and Kyoungsoo Ahn are researchers and **Jinhyoun Park** is an engineer at Hyundai Heavy

Industries (HHI). They are responsible for developing smart ship solution in Hyundai Heavy Industries. Their previous experience includes developing a vessel management system and analysing ship's performance at sea.

AUTOMATED/CONTROLLED STORAGE FOR AN EFFICIENT MBOM PROCESS IN THE SHIPBUILDING MANAGING THE IoT TECHNOLOGY

Arturo Benayas-Ayuso, Universidad Politécnica de Madrid/Spain, abenayas@alumnos.upm.es
Rodrigo Perez Fernandez, Universidad Politécnica de Madrid/Spain, rodrigo.perez.fernandez@upm.es

SUMMARY

Non automated/controlled warehouse sets a possible improvement point in several shipyards. Material arrives into shipyard in stock-trucks or ships, and it is randomly stored when a tight time is set. Material stock orders are saved in the Shipyard storage control System, but there are no link between orders and the real arrived material.
MBOM production working orders are based in material available in warehouses, new arriving material sets an availability to start processing this order, but sometimes, MBOM material preparation request more time than expected. Setting an arrival sorting and tag process, with some special marking methods could help to improve material control, availability and processing for these MBOMs. This also increases the material availability time in the Shipyard Storage control System, but it reduces material processing for the next steps of the ship production.
Another problem is that, in tight schedules, material collection could be too much, so extra material represents unnecessary purchases. Some material requires a three month request to be served to shipyard. This period gap sets a delay to adjust the real required material to the available in the shipyard, to avoid this delay impact, shipyards try to request in advance more material than real needs, in order workshops don't decrease their work-load.
Second problem must be resolved using the CAD and PLM Systems in order to help engineers generating a fast-close-to-real material collection prevision, using special pre-collection of information and helping designers to improve the 3D model.

NOMENCLATURE

BAP	Battery-Assisted Passive
CAD	Computer Aided Design
CDMA	Code Division Multiple Access
EBOM	Engineering Bill of Material
ERP	Enterprise Resource Planning
GSM	Global System for Mobile comm.
HF	High Frequency
IIoT	Industrial Internet of Things
IoT	Internet of Things
ITU	International Telecomm. Union
ISM	Industrial-Scientific-Medical
LF	Low frequency
MAC	Medium-Access Control
MBOM	Manufacturing Bill of Material
NFC	Near Field Communication
OS	Operating System
PJM	Phase-Jitter Module
PLM	Product Lifecycle Management
PSK	Phase-Shift Keying
P&ID	Piping & Instrumentation Diagram
RFID	Radio Frequency IDentification
SCS	Storage Control System
TDMA	Time Division Multiple Access
UHF	Ultra-High Frequency
UMTS	Universal Mobile Telecomm. System
WIA	Wireless Industrial Applications

1. A BRIEF INTRODUCTION TO THE IoT

The ship projects are developed with CAD platforms, but every day we are looking for integrated development of the product involving all its Life Cycle. *Computer Aided Design* (CAD) system integrated with *Product Lifecycle Management* (PLM) and from the PLM we can conceive all the design but also control the production and include the use of the vessel.

The PLM can contain information of all systems of the vessel and also all its components. If the components are designed for the *Internet of Things* (IoT) it will have technology that allows sharing their situation, diagnosis, functionality with the PLM system which distributes the initial design.

The PLM system can use this information for knowing whether they are working properly or if we can improve its performance. It is also possible to identify whether it is necessary to make maintenance of the object or if it is necessary to replace it because its life ends or because it's working wrongly. It will be possible to determine and evaluate its performance comparing to other similar components or comparing to its different operating periods. It will also be possible to know how their performance affects the functioning of the whole product, i.e., the vessel. Furthermore, if the connection of the objects is realized with its PLM, it would be possible to record their history status, make change tracking, and know what is its function or its performance after realizing programmed maintenance.

In case of a vessel, this connectivity will be extended to the commercial mission to act autonomously in operation conditions. A commercial vessel can transmit its navigation situation, load situation, the things to be discharged or to be recharged.

All these means a huge amount of information to be managed and analyzed. New programs have to be developed to obtain the best use of such information so that the design can be improved from real function information of the design and it can be self-maintained with the connection with this huge cloud information to create methods that the objects can achieve certain intelligence.

The growth of the IoT is linked to the increase of Information and the management of big data, with the property that somehow IoT identifies Information and direction and order to a specific purpose, while the concept of Big Data is more generic.

The possibilities are countless, but the beginning is the same. It must begin in the initial design. It is necessary to consider what is needed to correctly fulfill the mission of the atomic elements. These requirements must be configurable in the initial design from where it will be extended to relations between each of them with other entities. CAD is one of the first steps, because it is where begins to collect systematically the concept of each component. That is why it is necessary to provide CAD tools to carry out the design for IoT.

2. OUR VISION AS REGARDS SMART SHIPS AND CONNECTIVITY

Porter and Heppelmann [2] described the new technology stack: *"Smart, connected products require companies to build and support an entirely new technology infrastructure. This technology stack is made up of multiple layers, including new product hardware, embedded software, connectivity, a product cloud consisting of software running on remote servers, a suite of security tools, a gateway for external information sources, and integration with enterprise business systems"*. If a company wants enter in this business must have products that connects with the products of other companies. So, the necessity of having an opened product is increasing. Also, Taylor [1] affirmed *"The platform is the key to success. The things will get increasingly inexpensive, applications will multiply, and connectivity will cost pennies. The real value will be created in the horizontal platform that ties it all together, i.e. the new OS."*

From the authors' point of view, the mentioned platform it is not exactly an *Operating System* (OS) [3], but a set of interconnected applications, which share the same data about the product. Data that have been appearing since the early design phases and feed that single database, not only with the design attributes but also with the desirable values for optimal performance. Around this database will appear the applications that control the life cycle of the components, the sensors that collect the information of operation or operational life of the same objects and devices.

The key is that all devices, sensors, components are not connected directly to the IoT, but connected to a management core that controls the data to interchange to IoT for those devices allowed to enter in that network. In this way, it is possible to take advance of the fact that all the existing components in the ship have its respective model created, incorporated, and defined from the CAD System. We could say that CAD is the birthplace of the elements of the ship. Therefore, it is the CAD System that knows them and from where they can be assigned the mission and the level of participation in that world of IoT.

CAD tools are at the beginning of the product life cycle, but they also have a strong relationship with the production cycle, providing the information needed for construction in all aspects. Those CADs with a compact and homogeneous database can extend their contribution to both the life cycle management tools and the IoT connectivity management application of the ship's elements.

To make possible all of this, it is necessary that CAD applications have certain characteristics. A CAD System has to evolve to become a global solution. In our vision, the CAD System will not only be the heart of the design but also the vehicle of communication between the products, their manufacturers and their operation. Having this idea in mind, we wonder: what should be those characteristics that turn a CAD System into a global design solution? We think the answer is the integration of systems. It is impossible to grow enough quickly just with organic grew. So, the companies have to look for the future of their products thinking in terms of integration. The question now is which the way of getting the integration is. To answer this question, it is necessary to look into the future and to identify the technological trends that will be developed in this area.

We consider that the connectivity, artificial intelligence and virtual reality are the keys to designing the future in the next ten years. We also think about what will be the application of these trends in the naval field. Any future technology for software solutions, the world of IoT, Smart Ships and any other technological developments of the products shall be supported by some of these three trends. They are not disjointed roads, they are trends that converge, and support each other.

3. RFID IIoT SENSORS: DEFINITION AND CHARACTERISTICS

First of all, which sensors can be considered as an IIoT sensor? To answer this question, the most extended definitions can be sum-up as *"Any kind of sensor, which can transmit its data in a network, internal or external, can be considered as an IoT sensor"*. When this is applied to an industrial environment, these sensors become IIoT sensors.

Based on this, sensor can use, as data transfer technology: ultrasonic sound, infrared light, Wi-Fi, Bluetooth, ZigBee, cellular mobile communication as

Global System for Mobile communication (GSM) or *Universal Mobile Telecommunication System* (UMTS); ultra-wide-band, *Radio Frequency IDentification* (RFID) or any available standard method approved by the IoT Standard Committee [10].

RFID tag has been selected as main article sensor because: are cheap, and its price is decreasing with the extended production, can be waterproof and anti-metal, and can be read/write several times with no problem, making them reusable.

RFID sensors can be classified by frequency and powered/non-powered tags [6] [7] [11]. This is well-known but we consider that is necessary to mention it briefly to clarify the basic concepts related to the methodology and the hardware available, also, described characteristics affects to the selection of the corresponding RFID tag to be used in warehouse environments.

Sensors can be classified by Frequency as follows:

- *Low Frequency* (LF) RFID. According to the *International Telecommunications Union* (ITU), the LF band goes between 30 kHz and 300 kHz. Frequency and power in this band are not regulated globally in the same way: most systems operate at 125 kHz, but there are some at 134 kHz. The reading range provided is short (generally up to 10 cm), so, in practice, LF devices are not usually sensitive to radio interference. Its most popular applications are access control and animal identification (mainly for pets and livestock).
- *High Frequency* (HF) RFID. Although the HF band goes from 3 MHz to 30 MHz, most systems operate at 13.56 MHz. HF systems can reach a reading distance of up to 1 m, what can lead to interference and, consequently, *Medium-Access Control* (MAC) mechanisms are recommended. This sort of RFID systems is massively used in transportation, payment, ticketing, and access control.
- *Ultra-High Frequency* (UHF) RFID. The UHF band actually covers from 300 MHz to 3 GHz, but most systems operate in the *Industrial-Scientific-Medical* (ISM) bands around 860–960 MHz and 2.45 GHz. 5.8 GHz has been abandoned [11]. UHF tag can be easily read at 10 m, so they are ideal for inventory management and item tracking in logistics.

All these RFID sensors can also be classified according to the way the tags are powered:

- Passive systems. They do not need internal batteries to operate, since they rectify the energy sent through the reader's antenna. There are LF, HF, and UHF passive systems, which nowadays can be easily read at a 10-m distance.
- Active systems. They include batteries, what allow them to reach further distances (usually up to 100 m). Due to power regulations, almost all active systems operate in the UHF band.
- Semi-active, semi-passive or *Battery-Assisted Passive* (BAP) systems. They decrease power consumption by using batteries just for powering the tags for certain functionality. Commonly, batteries are used to power up the basic electronics, while the energy obtained from the reader is used for powering the communications interface.

The *Electronic Communications Committee* (ECC), in its last report [5], sets in the annexes 1, 2, 11 & 12 the recommended frequencies for RFID, and distinguish it from medical, annex 12, and industrial annex 11, the other two annexes are general recommendations about devices/sensors used.

Frequency has been divided to do not get interferences between closely sensors. Last standard applied [4], set in its Part 4, for *Industrial-Scientific-Medical* (ISM), a frequency of 2.45 GHz to work with industrial RFID tags.

As result of these two rules applied, in most of European countries, UHF is divided in:

- Industrial applications, 2446-2454 MHz.
- Medical applications, 2483.5-2500 MHz.

Actual RFID tags manufacturers offer for minimum failure of reading tags the 13.56 MHz HF, which includes the *Near Field Communication* (NFC) technology field. Passive tags based on this frequency, allows up to 3 m reading distance, which is also profitable for the application we are describing.

NFC which works also in this frequency, has a limited read field set lower than 4 centimeters. This technology is not suitable for shipbuilding application.

Another frequencies can be used, but for warehouse application it is recommended this set. This is explained better in next section.

4. SELECTING AN ADEQUATE SENSOR FOR SHIPBUILDING WAREHOUSE

Shipyard warehouses have a common characteristic, most of materials managed in are made of metal, mainly steel alloys, but also some aluminum alloys.

Any transmission made is affected by the metal reaction to electromagnetic waves causing interferences to the information submitted. This represents a big issue for sensor selection, depending on frequency, greater values generates growing interferences in quantity and magnitude, doing data unreadable [8].

Second issue is piled materials, like steel sheets, they require that on-read tag doesn't generate interferences with close tags, shadows or inactivation, generating a non-interpretable data, and also increasing material sort time or tag time in order not to set RFID labels so close to avoid this undesired data disruption [8].

Grouping these with other characteristics required for the RFID tags used in warehouses, the following list is obtained:

- Fast read.
- Fast print & coded in real time.
- Durable and waterproof.
- Complete information saved in.
- External readable information for human-operators.
- Easy implementation in common devices, like tablets or smartphones, to avoid a double investment in materials dedicated to this *Storage Control System* (SCS).
- Help to localize items in warehouses [10].
- Avoid interferences, own or with labeled-close-to materials.
- Due to piled material, multi-reading tags.

Most of RFID tags match with this characteristics, but one of the biggest issue, interferences and multi-reading close tags, like piled ones attached to metal sheets is not as easy for them.

Standard method for RFID tags reading is based on *Time Division Multiple Access* (TDMA), but this old method generates a huge set of collisions between close tags [9], creating interferences, shadows or, even, non-activated tags due to a loss of energy, see [Figure 1].

To solve the matter of interferences, depending on frequency, several methods are available:

- HF, it is solved by *Phase-Jitter Module* (PJM) passive tags [11].
- Same concept can be applied to UHF tags with certain chips which admits *Phase-Shift Keying* (PSK) modulation method in UHF, reducing all effects of interferences between tags.
- Another method is based on the *Code Division Multiple Access* (CDMA) is applied to BAP tags, allowing them to communicate [9].

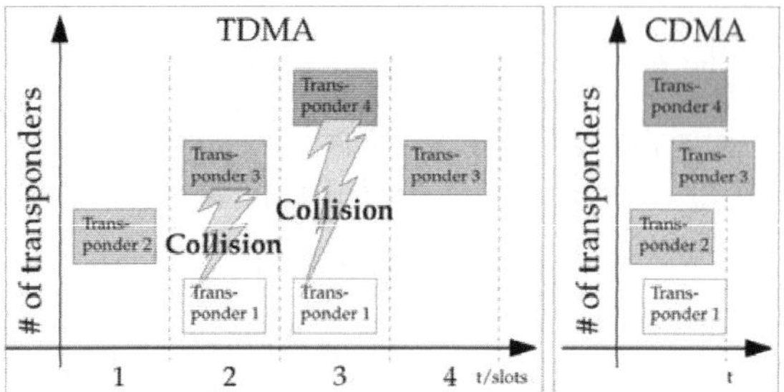

Figure 1. Comparison of TDMA and CDMA communication channel access techniques for RFID [9].

For anti-metal protection, several insulation materials have been used, such as FR4 (glass-reinforced epoxy laminate material) or other plastic ones, which indirectly confers an IP68 protection, making this tags waterproof, durable and isolated.

All this tags allows, depending on manufacturers, a printable cover that is filled at the same time the sensor is coded, reducing the available time.

Summing up all this characteristics:

- It can be read easily, only requires an antenna. Now a days, it can be read from phones or tablets, it is not required a dedicated or specific tools, which increases the capability of the system, allowing run at the same time an access to the SCS.
- It can be prepared, printed and coded in real time, reducing errors in material availability in the SCS, at the same time the tag is coded, material is uploaded to the system. This requires a specific tool.
- It can be combined with BAP tags for high important material, makes it easy to locate in warehouse.
- No optical readers are required, avoiding tag dirtiness, superficial damage or other visual modification.
- If a material is modified, standard plate scrap, if tag is not damage in the process, it can be recoded to match the new material assigned.
- Material inventory can be easily done with a multi-tag-reader, due to each tag has its own code. Also, using a control-door (tag-reader portal as used in standard shops), SCS can discount automatically the items from the inventory.

5. CONNECTING WAREHOUSE TO THE STORAGE CONTROL SYSTEMS, ERP/PLM AND CAD SYSTEMS

5.1 WAREHOUSE AND SCS

Warehouses in shipyard are specialized (steel work, pipe work, etc.), distributed (close to workshops and test centers) and should have a centralized SCS which, in a manner we are going to describe, should be connected with engineering.

Some hardware tools needs to be installed:

- A set of 3 or more inner warehouse antennas, which helps to control storage and locate material. This antennas should be only active in location process.
- Active doors can help to automatically add arriving material to the SCS or discounting it from available stock. All this is made after items are marked, inside or outside the shipyard.
- Material marked at vendors' place, if tags are compatible with the SCS hardware installed in our shipyard, can be directly added to our stock with no effort, helping warehouse control to count and locate easily.
- Active door and inner antennas combined can help to place arriving material in warehouse, and also memorize final location for fast recover of it.

- Manual user readers, if it cannot be implemented in tablets or smartphones used in warehouse, to control stock.
- Active balances for small fittings used in pipes, to count or discount valves and other elements from outfitting warehouse. This method can be useful also for cables.
- Printers-coders for RFID tags to be written in real-time and overwritten when material changes.

Traceability of materials in warehouses is made by the SCS with an active connection to the described hardware. These SCS software can be a module of the CAD system, delegated in an Enterprise Resource Planning (ERP) or in a PLM dedicated module, or a custom solution based in any of the previous platforms or, even, an isolated tailor-made service/product.

Depending on the selected software, some capabilities are available or require to be customized in the shipyard.

For this article, we assume that SCS selected software can:

- Control specialized material codes, such as used in RFID tags, with warehouse assignment, and internal location. Also can save the internal tag identification code which makes unique the label.
- Communicate with CAD system, to help engineering, allowing or denying material use, helping with material interchange, etc.
- Communicate with PLM or ERP, to help production:
  - Adding non engineered materials to orders *Engineering Bill of Material* (EBOM) → *Manufacturing Bill of Material* (MBOM) conversion.
  - Setting pre-collection material orders.
  - Working orders (MBOM).
  - End-of-work orders which generate a new material for next production step.

- Communicate with PLM or ERP, to help maintenance, setting collection orders for the full lifecycle of the ship. In this part, when a modification in ship is required, by a material change, also, some engineering improvement work should be done in the CAD System to have the digital model align with sailing ship. This digital maintenance model can be shared with ship owner.

Internal location can be triangulated depending on the antennas available in the warehouse, and location method precision changes depending on the tag type used [10], from meters in case of HF to centimeters in case of passive UHF, or up to 3 centimeters in the case of BAP UHF tags [10].

5.2 CAD SYSTEM AND SCS

Engineers design the ship in accordance with ship owner and classification society requirements, but some materials change, due to new regulations applied during ship design time, unavailability of materials or deprecated ones.

In all these cases, SCS software should help design team to identify easily the non-appropriate materials in the 3D modeling, reducing cost derived for this issue, highlighting them in the scene.

For pre-collection of materials in outfitting environment, intelligent tools connected to the SCS software can count the required materials, based on the intelligent *Piping & Instrumentation Diagrams* (P&ID) which uses the required materials in this 2D design.

These diagrams can have a two phase of design, basic and detailed, first one is applied in early steps of the project and represents main equipment requirements and accessories, but second one, applied in 2D or directly in 3D generates more accurate information.

Electrical and electronic systems are based on the same principles, and CAD should help designers to avoid, in details design, interferences.

In the case of main structure, this design is directly made in 3D in modern CAD softwares, preventing delays in steel work for the blocks and sub blocks in which is divided the whole ship.

In this case, pre-collection is done based in the standard shipyard scrap percentage and the required material, based on quality, thickness and surface required.

SCS software should have a bidirectional connection with the CAD system to feed the catalog components in, reducing designers awaiting time to use it, and when systems/structure are mature enough, feeding the SCS software with all used/required materials to emit the pre-collection orders.

Last part, maturity of information, is a team design leader decision which requires human interaction, and, based on CAD tools like design-rules and auto-checking, can be easily and fast done.

5.3 CAD, ERP, PLM AND SCS SYSTEMS INTEGRATION

Materials need to be collected, from CAD design by maturity, added to the ERP/PLM to Control the workflow in the workshops and counted in then SCS software to help people in purchasing department to emit the collection orders.

This is based on full interoperability between systems, working all of them like a unique software system.

Material in workshops can be also tagged, coded and storage for the next production step, helping CAD system to lock items making them non-editable, PLM/ERP advance in the corresponding workflow to activate next production step, and SCS software to add a new conformed-material to the production steps.

6. CONCLUSIONS

Future Smart Ships or Smart Shipyards must be connected to IoT or they will not be smart. And probably the cheaper option to achieve it is using RFID tags, due to its low cost and easy codification. They are an affordable solution for warehouse tracking system, helping shipyard with material traceability and lowering material collection times for MBOM orders.

This kind of tags can be used not only in warehouses to control stock also in workshops and on board, making really profitable the investment in all readers and writers.

Also, if software systems fails, due to these tags are out of the system, with a manual reader we can control the storage easily and make unstoppable production line, isolating it from other parts of the shipyard. And when software problem is solved, update SCS stock without any extra effort.

As a summary, just to mention some of the unique limits:

- Initial study of the requirements.
- Initial investment.
- And selection of a set of softwares which can fulfill with all requirements of efficiency for each group (CAD designers, production operators, purchase people...) and, at the same time, fully integrated with other part of this complex system.

7. ACKNOWLEDGEMENTS

The authors would like to express sincere gratitude to their respective organizations for providing the support and information for this paper.

8. REFERENCES

[1] S. Taylor, «10 Predictions for the Future of IoT» Big Data Quarterly, March 2016.

[2] M. E. Porter y J. E. Heppelmann, «How Smart, Connected Products are Transforming Companies» Harvard Business Review, October 2015.

[3] J.A. Muñoz and R. Pérez. «CAD tools for designing smart ships in the world of the internet of things (IoT)» SMART SHIP 2017, 24 and 25 January 2017.

[4] ISO (International Organization for Standardization) «ISO/IEC 18000-4:2015 Information technology - Radio frequency identification for item management - Part 4: Parameters for air interface communications at 2.45 GHz». Edition: February 2015.

[5] ECC (Electronic Communications Committee) «ERC Recommendation 70-03. Relating to the use of Short Range Devices (SRD)» Edition: October 2017.

[6] C.M. Roberts «Radio frequency identification (RFID)» Comput. Sec. 2006, 25, 18–26.

[7] T.M. Fernández-Caramés, P. Fraga-Lamas, M. Suárez-Albela and L. Castedo, «Reverse Engineering and Security Evaluation of Commercial Tags for RFID-Based IoT Applications» MDPI Sensors 2017, 17, 28.

[8] M. Roberti, «How can we minimize the failure rate of RFID?» RFID Journal, 2013-06-09.

[9] A. Loeffler, «Using CDMA as Anti-Collision Method for RFID - Research & Applications» InTech, DOI: 10.5772/16830. 2011.

[10] A. Loeffler and H. Gerhaeuser. «Localizing with Passive UHF RFID Tags Using Wideband Signals, Radio Frequency Identification from System to Applications» InTech, DOI: 10.5772/53769. 2013.

[11] ITU, «Analysis of Digital Data Technologies Toward Future Data Eco-Society» 2016

9. AUTHORS BIOGRAPHY

Arturo Benayas-Ayuso holds the current position of CAD & PLM Senior consultant at SENER Ingeniería y Sistemas S.A. He is PhD student at the Marine Engineering School of the Technical University of Madrid.

Rodrigo Perez Fernandez is the head of Naval Shipbuilding Business Development in SENER. He is professor in several subjects at the Marine Engineering School of the Technical University of Madrid.

AUTHORS' CONTACT DETAILS

Ian Kelsall
3M
ikelsall@mmm.com

Dan Hook
ASV Global
dan.hook@asvglobal.com

Arturo Benayas
Marine Engineering School of the Technical University of Madrid.
abenayas@alumnos.upm.es

Prabjot Singh Chopra
SeaPlus InfoNautics Pte Ltd
marketing@seatechsolutions.com

Rodrigo Perez
Marine Engineering School of the Technical University of Madrid.
rodrigo.perez.fernandez@upm.es

Seyong Jeong
Hyundai Heavy Industries Co. Ltd
seyong@hhi.co.kr

Yiannis Raptodimos
University of Strathclyde
yiannis.raptodimos.2013@uni.strath.ac.uk

Geraint R. West
Sonardyne International Limited
geraint.West@sonardyne.com

Natalie Mitchell
BAE Systems submarines
natalie.mitchell2@baesystems.com

Christos Gkerekos
University of Strathclyde
christos.gkerekos@strath.ac.uk

Michail Cheliotis
University of Strathclyde
michail.cheliotis@strath.ac.uk

Guang Zou
Lloyd's Register
Guang.Zou@lr.org

Munehiko Minoura
Osaka University
minoura@naoe.eng.osaka-u.ac.jp

Adrian Saw
Coltraco Ultrasonics
chloe@coltraco.co.uk